Biopesticides and Pest Management

Biopesticides and Pest Management

Vikram Sharma

RANDOM PUBLICATIONS
NEW DELHI (INDIA)

Biopesticides and Pest Management

ISBN 978-93-5111-200-6

Published in 2014 in India by

RANDOM PUBLICATIONS

4376-A/4B, Gali Murari Lal, Ansari Road
New Delhi-110 002
Phone : +91-11-43580356, +91-11-23289044
e-mail: randomexports@gmail.com, sales@randompublications.com,
info@randompublications.com

Reprinted 2025

Type Setting by : Keystoneprintads, Delhi-110051
Printed at : Replika Press Pvt. Ltd.

Preface

Biopesticides are certain types of pesticides derived from such natural materials as animals, plants, bacteria, and certain minerals. Biopesticides are a particular group of crop protection tools used in IPM. There is no formally agreed definition of a biopesticide. We define a biopesticide as a mass-produced agent manufactured from a living micro-organism or a natural product and sold for the control of plant pests. Examples of some biopesticides.

Biopesticides are typically microbial biological pest control agents that are applied in a manner similar to chemical pesticides. In order to implement these environmentally friendly pest control agents effectively, it can be important to pay attention to the way they are formulated and applied. Biopesticides for use against crop diseases have already established themselves on a variety of crops. For example, biopesticides already play an important role in controlling downy mildew diseases. Their benefits include: a 0-Day Pre-Harvest Interval (see: maximum residue limit), the ability to use under moderate to severe disease pressure, and the ability to use as a tank mix or in a rotational programme with other registered fungicides. Because some market studies estimate that as much as 20 per cent of global fungicide sales are directed at downy mildew diseases, the integration of biofungicides into grape production has substantial benefits in terms of extending the useful life of other fungicides, especially those in the reduced-risk category.

The first step in pest management is to identify the pest that is causing the problem. It then becomes essential to learn about the life cycle and behaviour of the pest to facilitate the development of a plan to manage it; the goal might be eradication or merely the reduction or suppression of its damage potential. A good pest management strategy incorporates some or all methods available to manage a given pest. This is called integrated pest management, or IPM. The goal should be to reduce pest populations and damage to economically and aesthetically tolerable levels. Complete eradication may not be possible, practical, or desirable. Integrated Pest Management (IPM) is a decision-making process that supports a balanced approach to managing crop and livestock production systems. The goal is effective, economical and environmentally-sound suppression of pests. For the purposes of this document, pests includes insects and mites, plant diseases, weeds, nematodes, and problem wildlife. The concept of IPM evolved in response to problems caused by an over-reliance on chemical pesticides. Some of these problems are development of pesticide resistance, elimination of natural enemies of pests, outbreaks of formerly suppressed pests, hazards to non-target species, and environmental contamination. The objective of IPM is not to eradicate a pest but to reduce its abundance to levels that no longer pose an economic threat to plants and animals.

Thus the book will be an immense help to researchers, environmentalists and agriculturists ensure an eco-friendly world.

I thank all members of my team who have helped in the preparation of the book. My special thanks go to "Random Publications" who have published the book.

– Vikram Sharma

Contents

1

Introduction

INTRODUCTION TO BIOPESTICIDES

Biopesticides are a particular group of crop protection tools used in IPM. There is no formally agreed definition of a biopesticide. We define a biopesticide as a mass-produced agent manufactured from a living micro-organism or a natural product and sold for the control of plant pests. Examples of some biopesticides. Biopesticides fall into three different types according to the active substance:

1. Micro-organisms;
2. Biochemicals; and
3. Semiochemicals.

The US Environmental Protection Agency also classes some transgenes as biopesticides

CLASSES OF BIOPESTICIDES

Biopesticides fall into three major classes:

1. Microbial pesticides which consist of bacteria, entomopathogenic fungi or viruses (and sometimes includes the metabolites that bacteria or fungi produce). Entomopathogenic nematodes are also often classed as microbial pesticides, even though they are multi-cellular.
2. Biochemical pesticides are naturally occurring substances that control (or monitor in the case of pheromones) pests, but are relatively non-toxic to mammals.
3. Plant-incorporated protectants (PIPs) have genetic material from other species incorporated into their genetic material (*i.e.* GM crops): still controversial in some (notably European) countries.

Biopesticides have usually no known function in photosynthesis, growth or other basic aspects of plant physiology; however, their biological activity against insect pests,nematodes, fungi and other organisms is well documented. Every plant species has developed a built-in unique chemical complex structure that protects it from pests. The plant kingdom offers a diverse array of complex chemical structures and almost every imaginable biological activity. These biodegradable, economical and renewable alternatives are used especially under organic farming systems.

Examples

Bacillus thuringiensis, a bacterial disease of Lepidoptera, Coleoptera and Diptera, is a well-known insecticide example. Because it has little effect on other organisms, it is considered more environmentally friendly than synthetic pesticides. The toxin from *B. thuringiensis* (Bt toxin) has been incorporated directly into plants through the use ofgenetic engineering.

Other microbial control agents include products based on:

- Entomopathogenic fungi (*e.g.Beauveria bassiana, Lecanicillium* spp., *Metarhizium* spp.),
- Plant disease control agents: include *Trichoderma* spp. and *Ampelomyces quisqualis* (a hyper-parasite of grape powdery mildew); *Bacillus subtilis* is also used to control plant pathogens.
- Beneficial nematodes attacking insect (*e.g. Steinernema feltiae*) or slug (*e.g. Phasmarhabditis hermaphrodita*) pests
- Entomopathogenic viruses (*e.g.. Cydia pomonella* granulovirus).
- Weeds and rodents have also been controlled with microbial agents.

Various naturally occurring materials, including fungal and plant extracts, have been described as biopesticides. Products in this category include:

- Insect pheromones and other semiochemicals
- Fermentation products such as Spinosad (a macro-cyclic lactone)
- Chitosan: a plant in the presence of this product will naturally induce systemic resistance (ISR) to allow the plant to defend itself against disease, pathogens and pests.
- Biopesticides may include natural plant-derived products, which include alkaloids, terpenoids, phenolics and other secondary chemicals. Certain vegetable oils such ascanola oil are known to have pesticidal properties. Products based on extracts of plants such as garlic have now been registered in the EU and elsewhere.
- Naturally occurring minerals such as baking soda may also have pesticidal applications.

Applications

Biopesticides are typically microbial biological pest control agents that are applied in a manner similar to chemical pesticides. In order to implement these environmentally friendly pest control agents effectively, it can be important to pay attention to the way they are formulated and applied. Biopesticides for use against crop diseases have already established themselves on a variety of crops. For example, biopesticides already play an important role in controlling downy mildew diseases. Their benefits include: a 0-Day

Pre-Harvest Interval, the ability to use under moderate to severe disease pressure, and the ability to use as a tank mix or in a rotational programme with other registered fungicides. Because some market studies estimate that as much as 20 per cent of global fungicide sales are directed at downy mildew diseases, the integration of biofungicides into grape production has substantial benefits in terms of extending the useful life of other fungicides, especially those in the reduced-risk category.

A major growth area for biopesticides is in the area of seed treatments and soil amendments. Fungicidal and biofungicidal seed treatments are used to control soil borne fungal pathogens that cause seed rots, damping-off, root rot and seedling blights. They can also be used to control internal seed–borne fungal pathogens as well as fungal pathogens that are on the surface of the seed. Many biofungicidal products also show capacities to stimulate plant host defences and other physiological processes that can make treated crops more resistant to a variety of biotic and abiotic stresses.

Advantages

- Harmful residues not detected
- Can be cheaper than chemical pesticides when locally produced.
- Can be more effective than chemical pesticides in the long-term (as demonstrated, for example, by the LUBILOSA Programme)

Disadvantages

- *High specificity:* Which may require an exact identification of the pest/pathogen and the use of multiple products to be used
- Often slow speed of action (thus making them unsuitable if a pest outbreak is an immediate threat to a crop)
- Often variable efficacy due to the influences of various biotic and abiotic factors (since biopesticides are usually living organisms, which bring about pest/pathogen control by multiplying within the target insect pest/pathogen)
- Living organisms evolve and increase their resistance to biological, chemical, physical or any other form of control. If the target population is not exterminated or rendered incapable of reproduction, the surviving population can acquire a tolerance of whatever pressures are brought to bear, resulting in an evolutionary arms race.

THE QUALITIES OF EFFECTIVE BIOPESTICIDES

The qualities of an effective biopesticide are similar to that of a chemical pesticide: they must work with good efficacy against the pest for which they are labeled. There has been considerable consumer and governmental pressure to minimize the use of chemical pesticides. This can be seen by the desire of the consumer to purchase organic produce and by supermarkets demanding that growers utilize environmental sensitive crop production strategies. The unique value for biopesticides is based on the three "R's":

Resistance Management

Since biopesticides have multi-modes of action, there is less chance of developing resistance in a particular insect. Biopesticides can be an excellent component of an integrated pest management (IPM) programme. An IPM programme involves a grower applying two or three chemical pesticides and then using a completely different product to vary the mechanism of action. Through government, supermarket and consumer pressure, IPM is becoming increasingly established as the future of crop protection.

Restricted Entry Intervals

Restricted entry intervals (REI's) determine the timing of application relative to a number of factors. The majority of biopesticides have low REI's, mostly around 0-4 hours with no pre-harvest interval. After a biopesticide is applied, a farmer can go into the field and harvest immediately.

Residues

Residues are a regulatory issue with pesticides. The Environmental Protection Agency (EPA) has jurisdiction over their proper application. The Office of Pesticide Programmes of the EPA has established a Biopesticide and Pollution Prevention Division (BPPD). This group helps to coordinate the registration of biopesticide products. Since biopesticides tend to pose fewer risks compared to conventional chemical pesticides, the EPA generally requires less data to achieve new product registrations (< 1 year). Conventional chemical pesticides can take up to three years for regulatory approval.

The 'risk cup' system places a limit on the total, cumulative, potential exposure to a pesticide across all of its applications. This system has forced companies to make the choice between registering specific agrochemicals solely for use on major crops, or to limit their use on major crops to enable their use on a larger number of less widely grown crops. Unsurprisingly, agrochemical companies have focused their registration of new pesticides on the high acreage field crops. Many have also taken the decision not to support older products through re-registration of high value niche crops. The USDA Inter-Regional Project No 4 (IR-4) programme is helping to support the registration of crop protection products on minor crops by carrying out some of the trials required for completion of a registration dossier, with the support of the relevant company. This government initiative has helped to limit some of the crop protection product availability problems faced by minor crop growers but has not entirely prevented a shortfall in crop protection product availability in some minor crop sectors.

The cost of registration of a biopesticide product is usually an order of magnitude less than for a chemical pesticide product. In the US, the cost of registration of a biopesticide can usually be measured in hundreds of thousands of dollars. In contrast, registration of a chemical pesticide costs millions of dollars. This reflects the regulator's preference for biopesticides, especially Bt products. Because of high chemical registration costs, many high value crop and niche markets are now being left without effective chemical crop protection solutions, leaving Bt as the economical choice for many producers. These markets represent a growth opportunity for InsectiGen™.

Additional Qualities of Biopesticides

Some other feature of biopesticides include: narrow target range, specific mode of action, slow-acting, safer than chemicals, limited field persistence, high unit cost of biopesticide production and they suppress but do not eliminate pests.

BIOPESTICIDE COMMERCIALIZATION

Worldwide there are about 1400 biopesticide products being sold. At present, there are 68 biopesticide active substances registered in the EU and 202 in the USA. The EU biopesticides consist of 34 microbials, 11 biochemicals and 23 semiochemicals, while the USA portfolio comprises 102 microbials, 52 biochemicals and 48 semiochemicals. To put this into context, these biopesticide products represent just 2.5 per cent of the total pesticide market. Marrone has estimated the biopesticides sector currently to have a 5 year compound annual growth rate of 16 per cent (compared with 3 per cent for synthetic pesticides), which is expected to produce a global market of $10 billion by 2017. However, the market may need to increase substantially more than this if biopesticides are to play a full role in reducing our overreliance on synthetic chemical pesticides.

Companies will only develop biopesticide products if there is profit in doing so. Similarly, the decision for a farmer whether or not to adopt a novel technology can be thought of in economic terms as a cost-benefit comparison of the profits to be made from using the novel versus the incumbent technology. A number of features of the agricultural economy make it difficult for companies to invest in developing new biopesticide products and, at the same time, make it hard for farmers to decide about adopting the new technology:

- *Lack of profit from niche market products.* Many biopesticides have high levels of selectivity. For example, bioinsecticides based on baculoviruses, such as the CpGV mentioned previously, typically are selective for just one or a few species of insect. This is of great benefit in terms of not harming other natural enemies and wildlife, but it means that biopesticides are niche market products with low profit potential. To quote Gelernter 'The features that made most Biological Control Products so attractive from the standpoint of environmental and human safety also acted to limit the number of markets in which they were effective'.

- *Fixed costs.* Because conventional chemical pesticides are used so widely, the fixed costs associated with them are spread over many users and hence represent a small part of the total cost of pest control. The knowledge needed by farmers to get effective control with pesticides is lower than with tactics such as biocontrol. Potential adopters of biopesticides face large fixed costs of adoption that will only decrease once the technology is used more widely, thereby disadvantaging early adopters.
- *Farmers' risk aversion.* For fruit and vegetable crops, cosmetic appearance is as important as yield when it comes to making a profit. The risks of producing an unmarketable crop are high, forcing growers to be risk averse with respect to new, untested crop protection technologies. Because conventional pesticides have been the mainstay of crop protection for over 50 years, there is a wealth of experience that gives farmers and growers confidence in their effectiveness. Farmers have achieved scale economies in pesticide use as a result of 'learning by doing'—the concept that one becomes more productive at a task the more it is repeated. In comparison, the more limited evidence base and practical experience with biologically based IPM technologies create uncertainty for farmers. Farmers' risk averse preferences can result in sub-optimal patterns of adoption of new technologies. Risk aversion is made worse if farmers' expectations of new technologies are more focused on the potential downsides rather than the benefits.
- *IPM portfolio economies.* Different IPM tactics work together as a 'technology bundle' or portfolio. If a farmer wants to switch from using a single chemical pesticide for pest control to IPM then (s)he will have to decide which combination of tactics to use. The number of potential portfolios to choose from increases rapidly as more tactics are included: with three tactics there are a total of seven different portfolios, with four tactics there are 11 different portfolios and so forth. Choosing the best portfolio in such cases is extremely challenging. The only realistic option is to develop a portfolio incrementally. Where a portfolio is already in place, then a farmer has to consider the benefits of adopting a new IPM tactic in the light of the current portfolio. Farmers want to use the minimum number of different tactics for the maximum benefit. Should the new tactic be added to the existing portfolio, or should it be used to replace an incumbent tactic? In some instances, it is

possible to replace a conventional synthetic chemical pesticide with a biopesticide without disturbing the existing IPM system (as in the case of using *B. bassiana* for control of spider mites in greenhouse IPM). In such a case, the new biopesticide technology can be adopted quickly and easily. However, IPM tactics may be synergistic, such that one tactic in the portfolio results in an improved performance in others. This is beneficial for IPM, but the interdependency of different tactics in this way can make it difficult to substitute with new technologies as they become available.

These factors mean that using conventional synthetic chemical pesticides applied on a calendar basis can be difficult to replace in favour of an IPM portfolio of alternative tactics including biopesticides. Chemical pest control may then become locked into the system until such a time that it fails, for example, if pesticide resistance becomes widespread, as in the greenhouse crops industry.

Pesticide 'lock in' also means that the adoption of new technologies will be biased towards tactics that closely resemble the incumbent pesticide technology. In the case of biopesticides, the products that have been most successful so far, such as microbial Bt, are very similar to chemical pesticides. This 'chemical model' of biopesticide development has encouraged companies to turn their attention away from the beneficial, biologically based characteristics of biopesticides (such as the ability of microbial agents to reproduce within host populations) and instead focus on trying to use biopesticides as chemical pesticide 'clones', resulting in unrealistic expectations of chemical-like efficacy.

It is important to stress that chemical pesticides are and will remain a vital part of crop protection. When used appropriately they can give excellent control with minimal adverse effects.

The use of chemical pesticides should therefore be promoted within an IPM framework so that they are used sparingly to minimize the evolution of resistance in target pest populations. However, IPM will only work if farmers have access to a range of crop protection tactics together with the knowledge on how to integrate them.

SYSTEMS APPROACH OF INTEGRATED PEST MANAGEMENT

There is an urgent requirement for alternative tactics to help make crop protection more sustainable. Many experts promote IPM as the best way forward, and the EU has placed it centrally within its 2009 Sustainable Use Directive on pesticides.

IPM is a systems approach that combines different crop protection practices with careful monitoring of pests and their natural enemies. The idea behind IPM is that combining different practices together overcomes the shortcomings of individual practices. The aim is not to eradicate pest populations but rather to manage them below levels that cause economic damage. The main IPM tactics include:

- Synthetic chemical pesticides that have high levels of selectivity and are classed by regulators as low-risk compounds, such as synthetic insect growth regulators.
- Crop cultivars bred with total or partial pest resistance.
- Cultivation practices, such as crop rotation, intercropping or undersowing.
- Physical methods, such as mechanical weeders.
- Natural products, such as semiochemicals or biocidal plant extracts.
- Biological control with natural enemies, including: predatory insects and mites, parasitoids, parasites and microbial pathogens used against invertebrate pests; microbial antagonists of plant pathogens and microbial pathogens of weeds.
- Decision support tools to inform farmers when it is economically beneficial to apply pesticides and other controls. These include the calculation of economic action thresholds, phenological models that forecast the timing of pest activity, and basic pest scouting. These tools can be used to move pesticide use away from routine calendar spraying to a supervised or targeted programme.

IPM can be done to different levels of sophistication. Prokopy outlines four levels: the basic level 1 combines different tactics against one pest on one crop; whereas the highest level 4 embraces all pests and crops on the farm within an overall Integrated Crop Management system that involves members of the broad policy network (extension services, industry, retailers, regulators) and takes account of the social, cultural and ecological context of farming.

An analysis of 62 IPM research and development projects in 26 countries, covering over 5 million farm households, showed that IPM leads to substantial reductions in pesticide applications. Over 60 per cent of the projects resulted in both a reduction in pesticide use (average reduction 75 per cent) and an increase in yields (average increase 40 per cent). Approximately 20 per cent of projects resulted in lowered pesticide use (average 60 per cent reduction) with a slight loss in yield (average 5 per cent reduction). Some 15 per cent of projects showed an increase of yield (average 45 per cent increase) with increased pesticide use (average 20 per cent increase); these were mainly conservation farming projects that incorporated zero tillage and therefore made greater use of herbicides for weed control.

The published evidence on the use of IPM by farmers outside of R&D projects is somewhat thin. For outdoor crops, IPM is based around targeted pesticide use, choice of cultivar and crop rotations. From a survey of 571 arable and mixed farms in the UK, Bailey *et al.* recorded reasonable levels of adoption of good pesticide practice, including use of seed treatments (*ca* 70 per cent adoption) and rotating pesticide classes (*ca* 55 per cent adoption), as well as good agronomic practice such as crop rotation (75 per cent adoption). However, adoption of more 'biologically based' IPM tactics was low, such as insect pheromones for pest monitoring (20 per cent) and introducing arthropod predators for biological control (7 per cent).

In contrast, biological control plays a central role in the production of many greenhouse crops. Pesticide resistance evolved in some key greenhouse pests as long ago as the 1960s, prompting the development of alternative methods of management. The pressure to reduce insecticide usage was reinforced by the adoption of bumble-bees within greenhouses for pollination. Some highly effective IPM programmes are now in place, based around the biocontrol of insect and mite pests using combinations of predators, parasitoids, parasitic nematodes and entomopathogens. Short-persistence pesticides are used on an at-need basis if they are compatible with biological control.

Pest management strategies are also determined through a close interaction between growers, consultants, biocontrol companies and retailers. In Europe, IPM based around biological control is used on over 90 per cent of greenhouse tomato, cucumber and sweet pepper production in The Netherlands and is standard practice for greenhouse crops in the UK. In Almeria, Spain, the area under biocontrol-based IPM has increased from just 250 ha in 2005 to around 7000 ha in 2008, while the proportion of the Dutch chrysanthemum crop grown under IPM increased from just 1 per cent in 2002 to 80 per cent in 2007 (R. GreatRex 2009, personal communication). This use of biological control requires considerable grower knowledge, but it has clear benefits in terms of reliable pest control, lack of phytotoxicity, a short harvest interval and better crop quality.

DECISION-MAKING PROCESS OF INTEGRATED PEST MANAGEMENT

Integrated Pest Management (IPM) is a decision-making process that supports a balanced approach to managing crop and livestock production systems. The goal is effective, economical and environmentally-sound suppression of pests. For the purposes of this document, pests includes insects and mites, plant diseases, weeds, nematodes, and problem wildlife. The concept of IPM evolved in response to problems caused by an over-reliance on chemical pesticides. Some of these problems are development of pesticide resistance, elimination of natural enemies of pests, outbreaks of formerly suppressed pests, hazards to non-target species, and environmental contamination. The objective of IPM is not to eradicate a pest but to reduce its abundance to levels that no longer pose an economic threat to plants and animals. IPM is not an organic or non-pesticide approach to pest control. Organic producers also use IPM and certain approved pesticides to protect their crops and livestock. One of the main goals of IPM is to promote the use of effective, less toxic pesticides only if and when necessary.

Elements of IPM

- *Pest Prevention*: Planning and managing agricultural production systems to prevent insects, plant diseases and weeds from becoming pests. Many pest prevention practices are cultural control practices. By preventing pest problems, there is no need for pesticides.
- Accurate identification of pests, their natural enemies and damage. Identification is essential to select the best pest management practice or pesticide for the pest/crop combination. It is illegal to use a registered pesticide on a crop or against a pest not listed on the pesticide label. Knowing what natural enemies are present and active in the crop will aid in selecting a pesticide that poses the least risk to them.
- Timely monitoring of pests and beneficial organisms, pest damage, and environmental conditions. Monitoring pest populations gives information needed to decide if and when to apply a pesticide for optimal effectiveness. Knowing what and when beneficial organisms are present is useful to time pesticide application to minimize impact on the beneficial organisms. Pest and disease development models use environmental (weather) information to determine what life stage(s) of a pest is present or if plants are at risk of disease infection. This information is useful in timing pest monitoring and management activities.
- Making control decisions based on pest or damage monitoring data, potential damage, cost of control methods, value of production, impact of other pests, beneficial organisms and the environment. Decide if the cost of the pesticide application is justified in terms of the value of the commodity protected. When making decisions consider the Economic Injury Level (EIL) and of Economic Threshold Level (ETL) where available. The EIL is the pest population level when the loss caused by the pest is equal to the cost of control measures. The ETL is the pest population or damage level when control measures are applied to keep the pest population from reaching the EIL. Both the ETL and EIL values can change from year to year depending on the crop value and control costs. For some pests the ET is zero because of quarantine regulations imposed by governments or markets.
- Using a combination of control methods - Use a combination of behavioural, biological, chemical, cultural and mechanical methods to reduce pest populations to acceptable levels.
  - Behavioural control applies to insects. It takes advantage of

insect responses to colours (*e.g.* yellow traps), odours (*e.g.* attractant-baited traps, sex pheromone dispensers for mating disruption), and light (*e.g.* black light traps, insect electrocutors).

- Biological control uses natural enemies to control pests. Examples of natural enemies (also called biocontrol or biological control agents) include predators (lady bugs and lacewings that feed on aphids), parasitic wasps and flies that attack caterpillars, nematodes that attack insects, insects that attack weeds, and diseases of insects (fungi, viruses) and weeds (fungi). Once established, biocontrol agents can often keep their hosts (pests) from becoming problems. If pesticide use is necessary, select products with no or little impact on biocontrol agents. This preserves their presence or allows the establishment of introduced biocontrol agents.
- Cultural control relates to production practices that discourage the introduction, establishment and development of pest populations. It minimizes the need for pesticides. Examples include growing pest resistant crop varieties, using disease-free plant materials, crop rotation, waste management, optimizing plant growth through proper use of plant nutrients, soil amendments, and sanitation.
- Mechanical control uses barriers or devices such as window screens, rodent traps, netting, fly paper, horticulture cloth, and mulches to exclude or destroy pests. These practices can reduce the need for pesticides.
- Chemical control uses pesticides to control pests. Often pesticides are used only when all other available control options have not kept pest populations below the ETL. New less toxic, more target-specific products (often referred to as reduced-risk pesticides) are replacing older broad-spectrum synthetic pesticides in order to reduce risks to food, environmental and human safety. However, these products on average cost more, and require more precise application timing and frequency.

• Evaluating the effects and efficacy of pest management actions. Use pesticide application and pest/damage monitoring records to assess how well the IPM programme worked. The records can help determine why the programme did not work as expected, and where improvements are necessary. Improvements may include better monitoring, use of more non-chemical pest management practices, or use of new reduced-risk pesticides.

In summary, pesticides will always be an important part of IPM programmes. IPM promotes careful use of pesticides based on an informed decision-making process. This process involves preventing pests and combining control practices based on current information on pest and beneficial presence, seasonal abundance, economic impact, and evaluation of programme performance.

Reduced Risk Pesticides

The Pest Management Regulatory Agency (PMRA) of Health Canada defines reduced risk pesticides as pesticides that may reasonably be expected to accomplish one or more of the following:

- Reduce the risks of pesticides to human health.
- Reduce the risks of pesticides to non-target organisms.
- Reduce the potential for contamination of groundwater, surface water or other valued environmental resources.
- Broaden the adoption of integrated pest management strategies, or make such strategies more available or more effective.

Reduced risk pesticides are either synthesized chemicals or chemicals derived from animals, plants, fungi or bacteria (called biopesticides). Examples of synthesized reduced risk pesticides are plant and insect growth regulators and chemicals that kill the pest but not the natural enemies.

Biopesticides include microbial pesticides (containing a live bacterium, fungus, virus, protozoan, or alga as the active ingredient), and chemicals derived from animals, bacteria, fungi and plants. A common biopesticide is*Bacillus thuringiensis*, or Bt, that has different strains to specifically control caterpillar, mosquito, or beetle pests.

REGULATORY BARRIERS TO BIOPESTICIDE COMMERCIALIZATION

Biopesticides encompass a very wide range of living and non-living entities that vary markedly in their basic properties, such as composition, mode of action, fate and behaviour in the environment and so forth. They are grouped together by governments for the purposes of regulating their authorization and use. These regulations are in place: firstly, to protect human and environmental safety; and secondly, to characterize products and thereby ensure that manufacturers supply biopesticides of consistent and reliable quality. The EU also requires that the efficacy of a biopesticide product is quantified and proved in order to support label claims. Only authorized biopesticide products can be used legally for crop protection.

The guidance of the OECD is that biopesticides should only be authorized if they pose minimal or zero risk. For example, the OECD guidance for microbial biopesticides is that: 'the micro-organism and its metabolites pose no concerns of pathogenicity or toxicity to mammals and other non-target organisms which will likely be exposed to the microbial product; the micro-organism does not produce a known genotoxin; all additives in the microbial manufacturing product and in end-use formulations are of low toxicity and suggest little potential for human health or environmental hazard'. The biopesticide registration data portfolio required by the regulator is normally a modified form of the one in place for conventional chemical pesticides and is used by the regulator to make a risk assessment. It includes information about mode of action, toxicological and eco-toxicological evaluations, host range testing and so forth.

This information is expensive for companies to produce and it can deter them from commercializing biopesticides, which are usually niche market products. Therefore, the challenge for the regulator is to have an appropriate system in place for biopesticides that ensures their safety and consistency but which does not inhibit commercialization. Until very recently, it is true to say that government regulators—with the probable exception of the USA—were unfamiliar with biologically based pest management and were therefore slow to appreciate the need to make the regulatory process appropriate for biopesticides rather than treat them in the same way as synthetic chemical pesticides. The decision whether or not to authorize a biopesticide product is made on the basis of expert opinion residing within the regulatory authority. When the regulators lack expertise with biopesticides, they tend to delay making a decision and may request the applicant to provide them with more data. There is also a risk that the regulator—using the chemical pesticide registration model—requests information that is not appropriate.

Some regulatory authorities, the UK, for example, have acknowledged that basing the regulatory system for biopesticides on a chemical pesticides model has been a barrier to biopesticide commercialization. A key question is whether the regulator, having recognized a problem, is able to do something about it. Social science theory indicates that government regulators and other bureaucratic organizations are vulnerable to 'goal displacement', during which they turn their focus away from achieving outcomes and instead concentrate more on internal processes. This can lead to systemic problems and stand in the way of introducing innovations into the regulatory system. This is not to say that regulatory innovation is not possible, and where there is sound evidence that a particular group of biopesticides presents minimal risk, the regulators have modified the data requirements. For example, the OECD regards semiochemicals used for arthropod control as presenting minimal hazard, with straight chain lepidopteran pheromones that form the majority of semiochemical-based biopesticides being thought sufficiently safe as to justify 'substantial reductions in health and environmental data requirements'. Other innovations are also being developed, which we discuss in the following sections.

PROMOTE BIOPESTICIDE USE

The EU passed a package of legislative measures in 2009 based around IPM, including the Framework Directive on the Sustainable Use of Pesticides (EU DG Environment). IPM principles do not become mandatory until 2014, but member states have been encouraged to use rural development programmes (funded under the Common Agricultural Policy) to provide financial incentives to farmers to start implementing IPM before this date. In the Commission's view, further research is still needed to develop successful crop-specific strategies for the deployment of IPM and this should include multidisciplinary research.

The Commission also regards it as 'crucial that Member States support the development of certified IPM advisory services organised by cropping systems to bridge the gap between research and end-users and help farmers for the adaptation of IPM principles to local situations'. Although such services can be provided privately and their quality guaranteed by a system of certification, it may be that countries that have retained state extension services, such as Denmark, have an inherent advantage in providing IPM advice in a cost-effective way.

Alongside the Sustainable Pest Management Directive, the EU also introduced a regulation that substantially amended the plant protection legislation embodied in Directive 91/414. This directive provided for a two-tier system of regulation involving the Community and member state levels. However, it quickly became evident that mutual recognition between different member states was not working, hence undermining the functioning of the EU internal market and deterring the development of biopesticides and other innovative products.

One of the solutions advanced was to divide Europe into climatically similar zones ('ecozones') where registration in one member state would facilitate registration in others in the same zone. This proposal proved controversial during the passage of the legislation. It was eventually achieved with northern, central and southern zones and an EU-wide one for greenhouses.

The new legislation gives a specific status to non-chemical and natural alternatives to conventional chemical pesticides and requires them to be given priority wherever possible. Biopesticides should generally qualify as low-risk active substances under the legislation. Low-risk substances are granted initial approval for 15 years rather than the standard 10. A reduced dossier can be submitted for low-risk substances but this has to include a demonstration of sufficient efficacy. One requirement for low-risk substances, that is still to be elaborated, is that their half-life in the soil should be less than 60 days and this may cause problems for some microbial biopesticides, such as rhizosphere-competent antagonists of soil-borne plant pathogens.

The new European legislation does not give the biopesticides industry all that it may have hoped for, but it does give biopesticides legislative recognition and opens up the potential for faster authorization processes and effective mutual recognition. This will require sustained work by those interested in the wider use of biopesticides. Many of the details of how mutual recognition in ecozones will operate in practice remain to be resolved, for example, how member states will interact with one another during the process. The achievement of real gains is very sensitive to the detailed implementation of the new procedures. What is clear is that the considerable variations in the levels of resource available to regulatory authorities in different member states will be a constraint on effective delivery.

STRATEGY OF PEST MANAGEMENT

The first step in pest management is to identify the pest that is causing the problem. It then becomes essential to learn about the life cycle and behaviour of the pest to facilitate the development of a plan to manage it; the goal might be eradication or merely the reduction or suppression of its damage potential.

A good pest management strategy incorporates some or all methods available to manage a given pest. This is called *integrated pest management, or IPM.* The goal should be to reduce pest populations and damage to economically and aesthetically tolerable levels. Complete eradication may not be possible, practical, or desirable.

Judicious selection of pest control methods should be aimed at reducing or eliminating pesticide use whenever possible. This is important because of concerns about personal and environmental safety. The following methods should be considered in developing a pest management strategy.

Integrated Pest Management relies on the combined use of physical, cultural, biological and/or chemical control of pests.

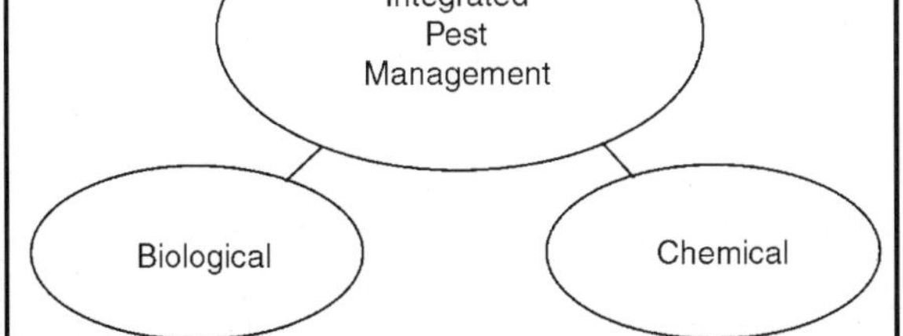

Physical control methods such as fly screens; physical means of proofing for birds, possums and rodents or physical means of pest detection, such as trapping, can sometimes be a more effective and appropriate means of pest control in gaining long term control over a particular pest infestation.

Cultural control methods of pest control such as improving ventilation to deter attack by termites or improving hygiene and sanitation measures to reduce the risk of cockroach infestation should always be undertaken, where possible, to make conditions less favourable for nuisance pests.

Biological control methods such as the introduction of parasites or predators to eradicate a particular pest have a growing importance in Integrated Pest Management. Chemical control of pests, while often necessary, should be undertaken judiciously using the appropriate pesticide and means of application to minimise any risk to human, animal or plant life.

PEST MANAGEMENT PRACTICES

The participating growers listed their top three weed, insect, and disease problems and indicated how they managed these problems.

Weed management was the most predictable hurdle that these growers faced. Growers employed three main strategies for weed control in vegetable production: hand weeding as needed (60 per cent), mechanical cultivation on an entire field (57 per cent), and crop rotations (46 per cent). Hand weeding was more common on smaller farms, while mechanical cultivation replaced this labour-intensive activity on larger farms.

Growers were more likely to use cultural practices in vegetables than in berries to reduce weed infestations and avoid future problems. Berries are a perennial crop, limiting cultural management options like crop rotations and adjusted planting dates for weed control. Most growers employed physical management practices in berries. For berries, hand weeding was the number one strategy (44 per cent), followed by using mulch (33 per cent) and mechanical field cultivation (22 per cent).

Only 17 per cent of the growers applied herbicides regularly to vegetables while 12 per cent did so in berries. Twenty-three per cent of the growers used herbicides only as needed in vegetables, and 12 per cent did so in berries. Diverse vegetable plantings limit herbicide use, as few herbicides are appropriate for application on multiple crops and carryover restrictions limit which crops can be planted 12 to 24 months after application.

Insect management becomes increasingly important as the size of the operation increases and fields become less diverse. Many growers took no action to manage insects (45 per cent in vegetables and 29 per cent in berries). Growers were likely to use insecticides only when risks or insect levels were high. Only 14 per cent of the growers used insecticides regularly in vegetables (5 per cent in berries). Cultural insect management practices in vegetables included crop rotations, crop residue incorporation, and adjusting planting dates and locations. Some berry growers used crop rotations, eco-friendly/natural insecticide applications, scouting, and plowing down crop residues.

Disease management is most often necessary when land has been in production for many years. Disease outbreaks are more likely to occur if growers do not employ crop rotations or use varieties not bred for disease resistance. Forty-seven per cent of the growers did nothing to manage vegetable diseases, while 26 per cent took no action to manage berry diseases. Crop rotations and disease resistant varieties were the top two disease management strategies. Twelve per cent of growers used regular fungicide applications to manage vegetable diseases, and 10 per cent did so for berry diseases. Growers also plowed down or removed diseased plant material, adjusted planting dates, and changed row spacings, density, or direction to control or prevent disease.

EXCLUSION

One of the safest and most effective ways to manage pests in the home environment is to deny them access. This is called exclusion, or pest-proofing, and may be accomplished via one of the following options.

Exclusion by Regulation

Government embargoes and quarantines which prohibit the introduction of pests into one country from another—or into one locality from another—are methods of exclusion by regulation. A case in point is the quarantine that prevents pine trees infested with European pine shoot beetles from being transported into Indiana counties known to be free of that insect.

Mechanical Exclusion

Barriers and devices such as fences, traps, lights, row covers, and noisemakers are examples of mechanical exclusion methods used to keep pests away from garden plants and out of homes. For example, nuisance wildlife such as rabbits can be excluded from gardens and landscape plantings with fencing. Some insects can be kept away from vegetables by covering them with a row cover made from a special kind of fabric. Birds can be banished from fruit crops by covering trees, bushes, or vines with plastic netting.

Rodents and other mammals such as bats can be excluded from homes by permanently closing entrance holes with caulking, steel wool, or structural repairs. Many insects can be kept out by caulking holes and cracks and ensuring that doors and windows are tightly sealed and screened. The use of yellow light bulbs (instead of the more conventional white bulbs) outside entrances will attract fewer kinds of insects. Storing food products and pet food in tightly sealed containers will guard against other insect and rodent infestations. Glue boards for cockroaches and traps for wildlife are other examples of devices that can be used to keep pests away from homes and plants. The force of water can be used to dislodge insects such as aphids from host plants; such slow-moving insects often simply die before they have a chance to crawl back onto the host.

One word of caution: Some exclusion devices—such as those that produce sounds audible only to insects or rodents, and light traps that electrocute insects—have not been proven effective.

CULTURAL CONTROL

Most plants and animals resist pests best when they are in good health. Therefore, keeping a potential host healthy can help prevent pest damage.

Plant Selection

It is important to select species and cultivars of crops and ornamental plants recommended for the locale. Neighbours and professionals in the community should be consulted to determine which kinds of plants grow best and without significant pest problems. Avoid those that are known to have a questionable history and those recognized as marginally hardy. Cold temperatures can predispose tender woody and perennial plants to pest damage; choosing a hardier variety would be an example of cultural control through plant selection. The European white birch tree is a beautiful tree and a popular choice for the residential landscape. However, it does not grow well in Indiana's climate; summer soil temperatures are too high, causing an unhealthy root system. The weakened tree then becomes susceptible to bronze birch borers that damage and usually kill it. Maintaining a European white birch tree in Indiana nearly always requires annual applications of an insecticide to control the borers. The frustration, expense, and hard work can be avoided by selecting a river birch, which is not susceptible to the borer, thereby eliminating the need for pesticide application. This, too, would constitute cultural control by plant selection.

Some plant cultivars resist or tolerate pest damage. Examples include tomato cultivars that are resistant to wilt diseases, apple cultivars resistant to scab diseases, and plants bred to produce more surface hairs that will discourage insect feeding. Cultivar selection should be based on the plants' known resistance to common pest problems, thus limiting loss potential and reducing the likelihood that a pesticide application might be needed.

Planting Dates

There are recommended planting intervals for most crops, and it is wise to recognize their importance. Careful selection of planting dates enhances crops' defences against disease and insect infestations. Planting too early in the spring can result in plants weakened by cold, wet soil conditions; and late spring frosts can damage or kill crops planted too early. Root and seed rots usually can be avoided by choosing later planting dates that lend more favourable soil conditions.

Certain insect pest problems can be avoided by choosing appropriate planting dates. Growing a combination of early, mid, and late season crops may decrease the potential for losses due to pests, based on the resulting staggered dates of maturity; *i.e.*, a pest present at a given time wouldn't be apt to affect all three stages of growth. Careful selection of planting dates is a form of cultural control.

Crop Rotation

If space permits, crops should be rotated to different areas of the garden each year to prevent buildup of pests in the soil.

Sanitation

Sanitation is perhaps the most important cultural practice that can be used to help manage pests. It consists of removing plants or plant parts suspected of harbouring insects or disease. For example, affected leaves, twigs, and branches of dogwoods infected with anthracnose should be removed and destroyed to help prevent the disease from spreading. Another example is the removal of certain plant parts that may be diseased, such as fallen rose leaves that are infected with black spot. It is important to always buy healthy seeds and plants known to be free of insects and disease. Examine 'gifts' from neighbours and tactfully decline those which obviously display pest symptoms. Remove garden weeds before they mature and produce seeds, and add them to your compost pile.

Sanitation is also important at the end of the growing season. Plant residue from annual crops, as well as the tops of herbaceous perennials, should be removed from the garden in the fall. Those not infested with insects or infected with disease can be added to a compost pile.

Other examples of cultural control through sanitation include removal of dead or diseased limbs from trees and shrubs; garbage management to discourage flies and rodents; careful attention to pet food areas; scrupulous cleanup of food crumbs in the home; and elimination of paper bags, newspapers, and other materials that provide food and shelter for pests such as cockroaches and rodents.

Other Cultural Methods

Good cultural practices include providing plants the best possible growing situation: proper spacing, watering, and fertilization. Weed control and the timely harvesting of produce also help to maintain healthy plants. The control of weeds with organic mulch is a good cultural practice, and mulch also contributes to plant health by moderating soil temperatures and conserving moisture.

Manipulation of a pest's environment also can be an effective method of cultural control. For example, venting the crawl space beneath a house will allow the space to dry, rendering the area unfavourable for the development of allergy-causing mildews and wood destroying organisms such as termites and decay fungi.

BIOLOGICAL CONTROL

Biological control utilizes the natural pathogens, predators, or parasites of a particular pest. One example is encouraging predatory lady beetles to destroy aphids in the garden. A biological approach might include growing certain plants as shelter or nectar sources for other predatory insects.

CHEMICAL CONTROL

Chemicals have been used for hundreds of years to control pests. Pesticides are chemical compounds formulated to control pests; roach sprays, weed killers, and rat and mouse baits are modern examples. Household and industrial cleaning products that kill germs are also pesticides, as are common bleaches as well as chemicals used to sterilize surgical tools and surfaces in hospitals. Pesticides can consist of one or more active ingredients, and the active ingredients can be either organic or inorganic. Organic compounds are based on carbon chemistry and are formulated from molecules that contain carbon, hydrogen, and oxygen. Some organic compounds used in pesticides occur naturally and are obtained from plants or bacteria; these are called natural, biological, or botanical pesticides. The organic active ingredient in pyrethrin insecticides is obtained from a certain chrysanthemum flower. Pyrethrins often are used to control flying and crawling insects in the home, as well as on pets and garden plants. Inorganic compounds do not contain carbon but are derived from mineral sources. Some inorganic compounds used in pesticides also occur naturally. An example is a copper-based product used for algae control in ponds.

Pesticides containing synthetic (manufactured) compounds comprise the largest number of products used to control pests. Most synthetic pesticides consist of organic compounds, and these represent the group that most people consider when contemplating the use of pesticides. Many synthetic pesticides have a mode of action similar to that of natural pesticides. Both natural and synthetic pesticides can vary in their toxicity to people and pets. Don't assume that just because a pesticide is 'natural' it is not toxic; conversely, some natural pesticides can be quite toxic. Always read the product label and follow the precautions stated. Follow all instructions carefully. Some pesticides control pests without killing them. Following are examples of pesticides that provide pest control alternatives.

Repellents

Topical products which lessen human attractiveness to ticks, chiggers, mosquitoes, biting flies, etc., are called repellents. They help make outdoor activities more enjoyable and contribute to the prevention of diseases transmitted by certain insects. Lumber treated with wood preservatives extends the lifetime of outdoor structures and furniture by repelling wood destroying organisms. Repellents can be used to discourage deer from feeding on the bark of valuable trees.

Attractants

Chemicals used to lure pests into a trap are called attractants. For instance, an insect sexual attractant (called a pheromone) may be used inside a plastic bag, luring Japanese beetles inside. When the bag becomes filled with beetles, it can be tied off and discarded.

However, Japanese beetle traps should be used with caution; since they work by *attracting* the beetles, one might end up with more beetles—instead of fewer—if they were the only person in the neighbourhood using the traps. Everyone in the neighbourhood should use them simultaneously to be effective; and it is important to place the traps far away from potential host plants.

Other examples of attractant devices are yellow jacket traps, glue boards for cockroaches, and traps used to monitor the presence of other insects.

Growth Regulators

Pesticides developed to adversely affect the growth of specific insects are categorized as growth regulators. They work by preventing the immature stages of certain insects from maturing into adults, or by rendering adults sterile or killing them. Some flea and roach products are growth regulators.

Desiccants

Insects that contact desiccants usually die. For example, a silica material called diatomaceous earth, commonly used in swimming pool filters and also helpful in managing garden slugs, is abrasive to the outer covering of some insects; the abrasive action dehydrates the pest, resulting in death.

Home Remedies

Some people try to formulate their own pesticides, using household products. This can be a dangerous practice. Many household chemicals are toxic, and mixing several of them together can result in combinations that are injurious to people, pets, and plants. This is especially true when the concoctions are applied to food crops. Remember that home remedies have not been tested and registered by the Environmental Protection Agency. Don't gamble with them!

DEVELOPING A STRATEGY OF PEST MANAGEMENT

When various pest management strategies—exclusion, cultural control, biological control, and/or chemical control—are used in conjunction with regular inspections of the home environment to monitor and manipulate pest activity, the process is recognized as integrated pest management; in most cases, an IPM approach is considered sensible and environmentally sound and will keep pest levels and damage below economically or aesthetically injurious levels. Although it is extremely important to remember that total eradication of a pest population is not the goal in most cases, it is equally important to recognize that sometimes it is. Eradication is by all means desirable when termites are damaging a structure or when pests present the possibility of disease transmission to people or pets.

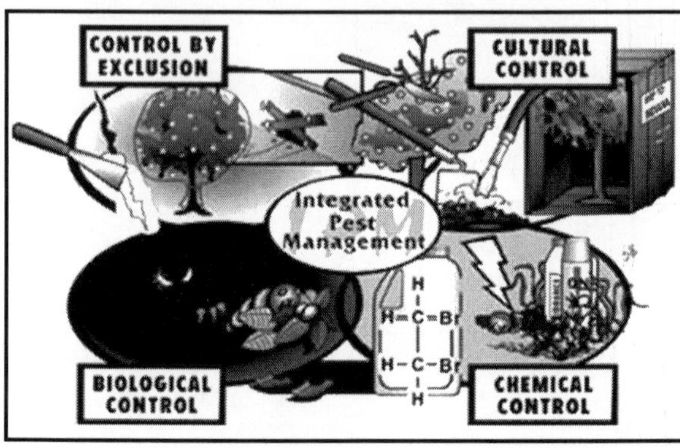

A preferred strategy in most pest control situations is to think in terms of reducing pest activity to a level which poses only minimal potential for damage to or annoyance of the host, be it plant, animal, or structure. For instance, it is not necessarily desirable to kill all spiders in the home environment; most of us are willing to tolerate a spider here and there, but we're not willing to share our kitchens with even a single German cockroach or mouse. Effective pest management and reduced use of pesticides are compatible. Pesticides used correctly and judiciously are good tools for pest control. However, they should not be used if non-pesticidal approaches can solve the problem. Selection of the most appropriate pest management method in a particular situation should be preceded by accurate pest identification and a survey of the site to determine the exact location and extent of the pest population. Pest management based on limited pesticide use requires learning as much as possible about the pest and how it survives in the home, lawn, or garden. In some cases, successful pest management strategies require patience, persistence, and long-term commitment.

Pest-Resistant Crops

One of the mainstays of integrated pest management is the use of crop varieties that are resistant or tolerant to insect pests and diseases. A resistant variety may be less preferred by the insect pest, adversely affect its normal development and survival, or the plant may tolerate the damage without an economic loss in yield or quality. Disease-resistant vegetables are widely used, whereas insect-resistant varieties are less common but nonetheless important. Examples include varieties of wheat which have tough stems that prevent development of the Hessian fly and cucurbits (squash, cucumbers, melons) that have lower concentrations of feeding stimulants (cucurbitacins) for cucumber beetles. Commercial corn lines with increased concentration of a chemical called DIMBOA display resistance to European corn borer, a major pest throughout the world. In the case of cabbage, the only reliable method of controlling onion thrips is through the use of resistant varieties.

Advantages of this tactic include ease of use, compatibility with other integrated pest management tactics, low cost, and cumulative impact on the pest (each subsequent generation of the pest is further reduced) with minimal environmental impact. The development of resistant or pest-tolerant plant varieties, however, may require considerable time and money, and resistance is not necessarily permanent. Just as insect populations have developed resistance to insecticides, populations of insects have developed that are now able to damage plant varieties that were previously resistant.

Cultural Control

There are many agricultural practices that make the environment less favourable to insect pests. Examples include cultivation of alternate hosts (*e.g.*, weeds), crop rotation, selection of planting sites, trap crops, and adjusting the timing of planting or harvest. Crop rotation, for example, is highly recommended for management of Colorado potato beetle. The beetles overwinter in or near potato fields and they require potato or related plants for food when they emerge in the spring. With cool temperatures and no suitable food, the beetles will only crawl and be unable to fly. Planting potatoes well away from the previous year's crop prevents access to needed food and the beetles will starve. The severity and incidence of many plant diseases can also be minimized by crop rotation, and selection of the planting site may affect the severity of insect infestations.

Trap crops are planted to attract and hold pest insects where they can be managed more efficiently and prevent or reduce their movement onto valuable crops. Early planted potatoes can act as a trap crop for Colorado potato beetles emerging in the spring. Since the early potatoes are the only food source available, the beetles will congregate on these plants where they can more easily be controlled. Adjusting the timing of planting or harvest is another cultural control technique. The earlier planted processing tomatoes grown in the western United States are far less likely to be infested by the tomato fruitworm than those planted later in the season. It is also important to use pest-free transplants. Some vegetable crop transplants can be infested with insect pests, and growers using these transplants are put at a considerable disadvantage.

2

Pesticides

Pesticides are used extensively to control pests, which can harm crops and other plant species. There are many different types of pesticides. They are mostly synthetic chemicals designed to affect a specific target. Concern over pesticide use is a major issue among many organizations. This chapter describes pesticides, why they are used, and why they generate so much concern. There are many plants and animals which co-exist with crops in a farmers field. Less than one per cent of these plants and animals are pests. Pests are organisms which attack crops and harm them in some way. It is critical that these pests are controlled for a farm to earn the best income. The use of synthetic pesticides in agriculture world-wide is still the most widespread method for pest control. Farmers could not achieve the same yields without the use of synthetic pesticides. It is estimated that without them 30 per cent of world crops would be destroyed by harmful organisms such as insects, weeds, or plant diseases even before they were harvested. Pesticides prevent these loses by controlling such pests. Pests can also render food unfit for consumption, harm human health and also harm the environment. In crops like cotton, tomatoes or melons, difficulties in pest control have forced farmers in some countries to abandon these crops completely.

DEFINITION

Food and Agriculture Organization (FAO) has defined the term of *pesticide* as: any substance or mixture of substances intended for preventing, destroying or controlling any pest, including vectors of human or animal disease, unwanted species of plants or animals causing harm during or otherwise interfering with the production, processing, storage, transport or marketing of food, agricultural commodities, wood and wood products or animal feedstuffs, or substances which may be administered to animals for the control of insects, arachnids or other pests in or on their bodies. The term includes substances intended for use as a plant growth regulator, defoliant, desiccant or agent for thinning fruit or preventing the premature fall of fruit. Also used as substances applied to crops either before or after harvest to protect the commodity from deterioration during storage and transport.

Type of Pesticide	Target Pest Group
Algicides or Algaecides	Algae
Avicides	Birds
Bactericides	Bacteria
Fungicides	Fungi and Oomycetes
Insecticides	Insects
Miticides or Acaricides	Mites
Molluscicides	Snails
Nematicides	Nematodes
Rodenticides	Rodents
Virucides	Viruses

Subclasses of pesticides include: herbicides, insecticides, fungicides, rodenticides, pediculicides, and biocides.

Pesticides can be classified by target organism, chemical structure, and physical state. Pesticides can also be classed as inorganic, synthetic, or biologicals (biopesticides), although the distinction can sometimes blur. Biopesticides include microbial pesticides and biochemical pesticides. Plant-derived pesticides, or "botanicals", have been developing quickly. These include the pyrethroids, rotenoids, nicotinoids, and a fourth group that includes strychnine and scilliroside.

Many pesticides can be grouped into chemical families. Prominent insecticide families include organochlorines, organophosphates, and carbamates. Organochlorine hydrocarbons (e.g. DDT) could be separated into dichlorodiphenylethanes, cyclodiene compounds, and other related compounds. They operate by disrupting the sodium/potassium balance of the nerve fiber, forcing the nerve to transmit continuously. Their toxicities vary greatly, but they have been phased out because of their persistence and potential to bioaccumulate. Organophosphate and carbamates largely replaced organochlorines.

Both operate through inhibiting the enzyme acetylcholinesterase, allowing acetylcholine to transfer nerve impulses indefinitely and causing a variety of symptoms such as weakness or paralysis. Organophosphates are quite toxic to vertebrates, and have in some cases been replaced by less toxic carbamates. Thiocarbamate and dithiocarbamates are subclasses of carbamates. Prominent families of herbicides include pheoxy and benzoic acid herbicides (e.g. 2,4-D), triazines (e.g. atrazine), ureas (e.g. diuron), and Chloroacetanilides (e.g. alachlor). Phenoxy compounds tend to selectively kill broadleaved weeds rather than grasses. The phenoxy and benzoic acid herbicides function similar to plant growth hormones, and grow cells without normal cell division, crushing the plants nutrient transport system. Triazines interfere with photsynthesis. Many commonly used pesticides are not included in these families, including glyphosate.

Pesticides can be classified based upon their biological mechanism function or application method. Most pesticides work by poisoning pests. A systemic pesticide moves inside a plant following absorption by the plant. With insecticides and most fungicides, this movement is usually upward (through the xylem) and outward. Increased efficiency may be a result. Systemic insecticides, which poison pollen and nectar in the flowers, may kill bees and other needed pollinators. In 2009, the development of a new class of fungicides called paldoxins was announced. These work by taking advantage of natural defense chemicals released by plants called phytoalexins, which fungi then detoxify using enzymes. The paldoxins inhibit the fungi's detoxification enzymes. They are believed to be safer and greener.

Uses

Pesticides are used to control organisms that are considered to be harmful. For example, they are used to kill mosquitoes that can transmit potentially deadly diseases like west nile virus, yellow fever, and malaria. They can also kill bees, wasps or ants that can cause allergic reactions. Insecticides can protect animals from illnesses that can be caused by parasites such as fleas. Pesticides can prevent sickness in humans that could be caused by moldy food or diseased produce. Herbicides can be used to clear roadside weeds, trees and brush. They can also kill invasive weeds that may cause environmental damage. Herbicides are commonly applied in ponds and lakes to control algae and plants such as water grasses that can interfere with activities like swimming and fishing and cause the water to look or smell unpleasant. Uncontrolled pests such as termites and mould can damage structures such as houses. Pesticides are used in grocery stores and food storage facilities to manage rodents and insects that infest food such as grain. Each use of a pesticide carries some associated risk. Proper pesticide use decreases these associated risks to a level deemed acceptable by pesticide regulatory agencies such as the United States Environmental Protection Agency (EPA) and the Pest Management Regulatory Agency (PMRA) of Canada.

Pesticides can save farmers' money by preventing crop losses to insects and other pests; in the U.S., farmers get an estimated fourfold return on money they spend on pesticides. One study found that not using pesticides reduced crop yields by about 10 per cent. Another study, conducted in 1999, found that a ban on pesticides in the United States may result in a rise of food prices, loss of jobs, and an increase in world hunger.

DDT, sprayed on the walls of houses, is an organochloride that has been used to fight malaria since the 1950s. Recent policy statements by the World Health Organization have given stronger support to this approach. Dr. Arata Kochi, WHO's malaria chief, said, "One of the best tools we have against malaria is indoor residual house spraying. Of the dozen insecticides WHO has approved as safe for house spraying, the most effective is DDT." However, since then, an

October 2007 study has linked breast cancer from exposure to DDT prior to puberty. Poisoning may also occur due to use of DDT and other chlorinated hydrocarbons by entering the human food chain when animal tissues are affected. Symptoms include nervous excitement, tremors, convulsions or death. Scientists estimate that DDT and other chemicals in the organophosphate class of pesticides have saved 7 million human lives since 1945 by preventing the transmission of diseases such as malaria, bubonic plague, sleeping sickness, and typhus. However, DDT use is not always effective, as resistance to DDT was identified in Africa as early as 1955, and by 1972 nineteen species of mosquito worldwide were resistant to DDT. A study for the World Health Organization in 2000 from Vietnam established that non-DDT malaria controls were significantly more effective than DDT use. The ecological effect of DDT on organisms is an example of bioaccumulation.

Amounts

In 2006 and 2008, the world used approximately 5.2 billion pounds of pesticides with herbicides constituting the majority of the world pesticide use at 40 per cent followed by insecticides and fungicides with totals of 17 per cent and 10 per cent respectively. The U.S. in 2006 and 2007 used approximately 1.1 billion pounds of pesticides accounting for 22 per cent of the world total. For conventional pesticides which are used in the agricultural sector as well in industry, commercial, governmental and the home and garden sectors, the U.S. used at total of 857 million pounds, with the agricultural sector accounting for 80 per cent of the conventional pesticide use total. Pesticides are also found in majority of U.S. households with 78 million out of the 105.5 million households indicating that they use some form of pesticide. Currently,there are more than 1,055 active ingredients registered as pesticides, which are put together to produce over 16,000 pesticide products that are being marketed in the United States.

Costs

On the cost side of pesticide use there can be a cost to the environment and human health, as well as the cost of the development and research of new pesticides.

Health effects

Pesticides may cause acute and delayed health effects in those who are exposed. Pesticide exposure can cause a variety of adverse health effects. These effects can range from simple irritation of the skin and eyes to more severe effects such as affecting the nervous system, mimicking hormones causing reproductive problems, and also causing cancer. A 2007 systematic review found that "most studies on non-Hodgkin lymphoma and leukemia showed positive associations with pesticide exposure" and thus concluded that cosmetic use of pesticides should be decreased. Strong evidence also exists for other negative outcomes from pesticide exposure including neurological, birth defects, fetal death, and neurodevelopmental disorder.

The American Medical Association recommends limiting exposure to pesticides and using safer alternatives: "Particular uncertainty exists regarding the long-term effects of low-dose pesticide exposures. Current surveillance systems are inadequate to characterize potential exposure problems related either to pesticide usage or pesticide-related illnesses...Considering these data gaps, it is prudent...to limit pesticide exposures...and to use the least toxic chemical pesticide or non-chemical alternative."

The World Health Organization and the UN Environment Programme estimate that each year, 3 million workers in agriculture in the developing world experience severe poisoning from pesticides, about 18,000 of whom die. According to one study, as many as 25 million workers in developing countries may suffer mild pesticide poisoning yearly.

One study found pesticide self-poisoning the method of choice in one third of suicides worldwide, and recommended, among other things, more restrictions on the types of pesticides that are most harmful to humans.

Environmental effect

Pesticide use raises a number of environmental concerns. Over 98 per cent of sprayed insecticides and 95 per cent of herbicides reach a destination other than their target species, including non-target species, air, water and soil. Pesticide drift occurs when pesticides suspended in the air as particles are carried by wind to other areas, potentially contaminating them. Pesticides are one of the causes of water pollution, and some pesticides are persistent organic pollutants and contribute to soil contamination.

In addition, pesticide use reduces biodiversity, reduces nitrogen fixation, contributes to pollinator decline, destroys habitat (especially for birds), and threatens endangered species. Pests can develop a resistance to the pesticide (pesticide resistance), necessitating a new pesticide. Alternatively a greater dose of the pesticide can be used to counteract the resistance, although this will cause a worsening of the ambient pollution problem.

Economics

Harm	Annual US Cost
Public Health	$1.1 billion
Pesticide Resistance in Pest	$1.5 billion
Crop Losses Caused by Pesticides	$1.4 billion
Bird Losses due to Pesticides	$2.2 billion
Groundwater Contamination	$2.0 billion
Other Costs	$1.4 billion
Total Costs	$9.6 billion

Human health and environmental cost from pesticides in the United States is a total of $9.6 billion:

Additional cost includes the registration process and the cost of purchase pesticides.

The registration process can take several years to complete the 70 different types of field test and can cost between $50–70 million for a single pesticide. Annually the United States spends $10 billion on pesticides.

Benefits

There are two levels of benefits for pesticide use, primary and secondary. Primary benefits are direct gains from the use of pesticides and secondary benefits are effects that are more long-term.

Primary Benefits

1. Controlling pests and plant disease vectors
   - Improved crop/livestock yields
   - Improved crop/livestock quality
   - Invasive species controlled
2. Controlling human/livestock disease vectors and nuisance organisms
   - Human lives saved and suffering reduced
   - Animal lives saved and suffering reduced
   - Diseases contained geographically
3. Prevent of control organisms that harm other human activities and structures
   - Drivers view unobstructed
   - Tree/brush/leaf hazards prevented
   - Wooden structures protected

Secondary Benefits

1. Community benefits
   - Farm and agribusiness revenues
   - Nutrition and health improved
   - Food safety and security
2. National benefits
   - Workforce productivity increased
   - Increased export revenues
   - National agriculture economy
3. Global benefits
   - Assured safe and diverse food supply
   - Less greenhouse gas
   - Reduced civil unrest

Monetary

For every dollar ($1) that is spent on pesticides for crops yields four dollars ($4) in crops saved. This means based on the amount of money spent per year on pesticides, $10 billion, that there is an additional $40 billion savings in crop that would be lost due to damage by insects and weeds. Generally speaking, farmers benefit from having an increase crop yield and from being able to grow a variety of crops throughout the year. Consumers of agricultural products also benefit from being able to afford the vast quantities of produce available year round. The general public also benefits from the use of pesticides for the control of insect-borne diseases and illnesses, such as malaria. The use of pesticides creates a large job market, which provides jobs for all of the people who work within the industry.

Alternatives

Alternatives to pesticides are available and include methods of cultivation, use of biological pest controls (such as pheromones and microbial pesticides), genetic engineering, and methods of interfering with insect breeding. Application of composted yard waste has also been used as a way of controlling pests. These methods are becoming increasingly popular and often are safer than traditional chemical pesticides. In addition, EPA is registering reduced-risk conventional pesticides in increasing numbers.

Cultivation practices include polyculture (growing multiple types of plants), crop rotation, planting crops in areas where the pests that damage them do not live, timing planting according to when pests will be least problematic, and use of trap crops that attract pests away from the real crop. In the U.S., farmers have had success controlling insects by spraying with hot water at a cost that is about the same as pesticide spraying.

Release of other organisms that fight the pest is another example of an alternative to pesticide use. These organisms can include natural predators or parasites of the pests. Biological pesticides based on entomopathogenic fungi, bacteria and viruses cause disease in the pest species can also be used.

Interfering with insects' reproduction can be accomplished by sterilizing males of the target species and releasing them, so that they mate with females but do not produce offspring. This technique was first used on the screwworm fly in 1958 and has since been used with the medfly, the tsetse fly, and the gypsy moth. However, this can be a costly, time consuming approach that only works on some types of insects.

Another alternative to pesticides is the thermal treatment of soil through steam. Soil steaming kills pest and increases soil health.

In India, traditional pest control methods include using Panchakavya, the "mixture of five products." The method has recently experienced a resurgence in popularity due in part to use by the organic farming community.

Push Pull Strategy

The term "push-pull" was established in 1987 as an approach for integrated pest management (IPM). This strategy uses a mixture of behavior-modifying stimuli to manipulate the distribution and abundance of insects. "Push" means the insects are repelled or deterred away from whatever resource that is being protected. "Pull" means that certain stimuli (semiochemical stimuli, pheromones, food additives, visual stimuli, genetically altered plants, etc.) are used to attract pests to trap crops where they will be killed There are numerous different components involved in order to implement a Push-Pull Strategy in IPM. Many case studies testing the effectiveness of the push-pull approach have been done across the world. The most successful push-pull strategy was developed in Africa for subsistence farming. Another successful case study was performed on the control of *Helicoverpa* in cotton crops in Australia. In Europe, the Middle East, and the United States, push-pull strategies were successfully used in the controlling of *Sitona lineatus* in bean fields. Plus many more cases where this strategy was more beneficial than simply using pesticides on their crops. Some advantages of using the push-pull method are less use of chemical or biological materials and better protection against insect habituation to this control method. Some disadvantages of the push-pull strategy is that if there is a lack of appropriate knowledge of behavioral and chemical ecology of the host-pest interactions then this method becomes unreliable. Furthermore, because the push-pull method is not a very popular method of IPM operational and registration costs are higher.

Effectiveness

Some evidence shows that alternatives to pesticides can be equally effective as the use of chemicals. For example, Sweden has halved its use of pesticides with hardly any reduction in crops. In Indonesia, farmers have reduced pesticide use on rice fields by 65 per cent and experienced a 15 per cent crop increase. A study of Maize yields in northern Florida found that the application of composted yard waste with high carbon to nitrogen ratio to agricultural fields was highly effective at reducing the population of plant-parasitic nematodes and increasing crop yield, with yield increases ranging from 10 per cent to 212 per cent; the observed effects were long-term, often not appearing until the third season of the study.

However, pesticide resistance is increasing. In the 1940s, U.S. farmers lost only 7 per cent of their crops to pests. Since the 1980s, loss has increased to 13 per cent, even though more pesticides are being used. Between 500 and 1,000 insect and weed species have developed pesticide resistance since 1945.

Regulation

International

In Europe, recent EU legislation has been approved banning the use of highly toxic pesticides including those that are carcinogenic, mutagenic or toxic to reproduction, those that are endocrine-disrupting, and those that are persistent, bioaccumulative and toxic (PBT) or very persistent and very bioaccumulative (vPvB). Measures were approved to improve the general safety of pesticides across all EU member states.

Though pesticide regulations differ from country to country, pesticides and products on which they were used are traded across international borders. To deal with inconsistencies in regulations among countries, delegates to a conference of the United Nations Food and Agriculture Organization adopted an International Code of Conduct on the Distribution and Use of Pesticides in 1985 to create voluntary standards of pesticide regulation for different countries. The Code was updated in 1998 and 2002. The FAO claims that the code has raised awareness about pesticide hazards and decreased the number of countries without restrictions on pesticide use.

Three other efforts to improve regulation of international pesticide trade are the United Nations London Guidelines for the Exchange of Information on Chemicals in International Trade and the United Nations Codex Alimentarius Commission. The former seeks to implement procedures for ensuring that prior informed consent exists between countries buying and selling pesticides, while the latter seeks to create uniform standards for maximum levels of pesticide residues among participating countries. Both initiatives operate on a voluntary basis.

Pesticide safety education and pesticide applicator regulation are designed to protect the public from pesticide misuse, but do not eliminate all misuse. Reducing the use of pesticides and choosing less toxic pesticides may reduce risks placed on society and the environment from pesticide use. Integrated pest management, the use of multiple approaches to control pests, is becoming widespread and has been used with success in countries such as Indonesia, China, Bangladesh, the U.S., Australia, and Mexico. IPM attempts to recognize the more widespread impacts of an action on an ecosystem, so that natural balances are not upset. New pesticides are being developed, including biological and botanical derivatives and alternatives that are thought to reduce health and environmental risks. In addition, applicators are being encouraged to consider alternative controls and adopt methods that reduce the use of chemical pesticides. Pesticides can be created that are targeted to a specific pest's life cycle, which can be environmentally more friendly. For example, potato cyst nematodes emerge from their protective cysts in response to a chemical excreted by potatoes; they feed on the potatoes and damage the crop. A similar chemical can be applied to fields early, before the potatoes are planted, causing the nematodes to emerge early and starve in the absence of potatoes.

United States

In most countries, pesticides must be approved for sale and use by a government agency. In the United States, the Environmental Protection Agency (EPA) is responsible for regulating pesticides under the Federal Insecticide, Fungicide, and Rodenticide Act (FIFRA) and the Food Quality Protection Act (FQPA). Complex and costly studies must be conducted to indicate whether the material is safe to use and effective against the intended pest. The EPA regulates pesticides to ensure that these products do not pose adverse effects to humans or the environment. Pesticides produced before November 1984 continue to be reassessed in order to meet the current scientific and regulatory standards. All registered pesticides are reviewed every 15 years to ensure they meet the proper standards. During the registration process, a label is created. The label contains directions for proper use of the material. Based on acute toxicity, pesticides are assigned to a Toxicity Class.

Some pesticides are considered too hazardous for sale to the general public and are designated restricted use pesticides. Only certified applicators, who have passed an exam, may purchase or supervise the application of restricted use pesticides. Records of sales and use are required to be maintained and may be audited by government agencies charged with the enforcement of pesticide regulations.

The EPA regulates pesticides under two under main acts, both of which were amended by the Food Quality Protection Act of 1996. In addition to the EPA, the United States Department of Agriculture (USDA) and the United States Food and Drug Administration (FDA) set standards for the level of pesticide residue that is allowed on or in crops The EPA looks at what the potential human health and environmental effects might be associated with the use of the pesticide.

Additionally, the U.S. EPA uses the National Research Council's four-step process for human health risk assessment:

1. Hazard Identification,
2. Dose-Response Assessment,
3. Exposure Assessment,
4. Risk Characterization.

History

Since before 2000 BC, humans have utilized pesticides to protect their crops. The first known pesticide was elemental sulfur dusting used in ancient Sumer about 4,500 years ago in ancient Mesopotamia. By the 15th century, toxic chemicals such as arsenic, mercury and lead were being applied to crops to kill pests. In the 17th century, nicotine sulfate was extracted from tobacco leaves for use as an insecticide. The 19th century saw the introduction of two more natural pesticides, pyrethrum, which is derived from chrysanthemums, and rotenone, which is derived from the roots of tropical vegetables. Until the 1950s, arsenic-based pesticides were dominant. Paul Müller discovered that DDT was a very effective insecticide. Organochlorines such as DDT were dominant, but they were replaced in the U.S. by organophosphates and carbamates by 1975. Since then, pyrethrin compounds have become the dominant insecticide. Herbicides became common in the 1960s, led by "triazine and other nitrogen-based compounds, carboxylic acids such as 2,4-dichlorophenoxyacetic acid, and glyphosate".

The first legislation providing federal authority for regulating pesticides was enacted in 1910; however, decades later during the 1940s manufacturers began to produce large amounts of synthetic pesticides and their use became widespread.

Some sources consider the 1940s and 1950s to have been the start of the "pesticide era." Although the U.S. Environmental Protection Agency was established in 1970 and amendments to the pesticide law in 1972, pesticide use has increased 50-fold since 1950 and 2.3 million tonnes (2.5 million short tons) of industrial pesticides are now used each year. Seventy-five per cent of all pesticides in the world are used in developed countries, but use in developing countries is increasing. In 2001 the EPA stopped reporting yearly pesticide use statistics. A study of USA pesticide use trends through 1997 was published in 2003 by the National Science Foundation's Center for Integrated Pest Management.

In the 1960s, it was discovered that DDT was preventing many fish-eating birds from reproducing, which was a serious threat to biodiversity. Rachel Carson wrote the best-selling book *Silent Spring* about biological magnification. The agricultural use of DDT is now banned under the Stockholm Convention on Persistent Organic Pollutants, but it is still used in some developing nations to prevent malaria and other tropical diseases by spraying on interior walls to kill or repel mosquitoes.

USE OF PESTICIDES TO REDUCE PEST POPULATIONS

Chemical control is the use of pesticides to reduce pest populations. The three major pesticide categories are herbicides, insecticides, and fungicides. Herbicides are chemicals that are used to control weeds, insecticides are chemicals used to control insects and fungicides are chemicals used to control plant diseases caused by fungi. There are many different types of these pesticides which affect pests in many different ways. Within each pesticide group are numerous chemicals for use in specific applications.

Depending on the chemical structure, the pesticide may interfere with the growth, reproduction or development of the pest or may induce damage to the internal systems, resulting in death. Some pesticides are more popular than others for reasons such as potency, availability, cost, and safety. There are usually a few brand names of pesticides which dominate the market. It is important to remember that no pesticides kill 100 per cent of the pests all the time. They simply kill enough pests so that the few remaining are no longer a problem. Most pesticides commonly used for agricultural, home and garden use belong to chemical families that are not toxic to humans or animals, do not pose cancer risk to humans and do not accumulate in the environment.

FARMERS SPRAY PESTICIDES

Application

When farmers spray pesticides, they try to distribute a small amount of active ingredient to the appropriate biological target without contaminating other areas. There are many different types of targets for pesticides. These include insects, plants, or soil. A variety of application techniques are required to affect these targets while using as little pesticide as possible. The equipment used for pesticide application ranges in price and complexity.

Some application techniques include:

- Release from the air in the solid state as dusts or granules or in the liquid state as sprays
- Injection into the plant
- Injection into the soil
- Release into water which is then used for irrigation
- Release from aircrafts when the areas to be treated are large and when operations on the ground are either not possible or will cause damage

Not Just on the Farm

Pesticides are more widely known for their use on farms to protect crops. Chemicals are used by people everyday to get rid of pests. Remember the last time you used Raid or Off? Pesticides are used mostly to control insects, in homes, office buildings, schools, hospitals, supermarkets, and many other common areas.

Most of these products are in the form of aerosols, pest strips, baits, pet products, and insect repellents. Pesticides are also used on lawns, gardens, golf courses, parks, and recreational areas.

PESTICIDE SAFETY

Because pesticides can kill pests, they also pose a potential risk to other organisms such as humans. For this reason, they are not allowed to be used without the most thorough checks and safeguards to ensure that they do not harm humans, other species, or the environment. Because of the concerns, synthetic pesticides have undergone a development process to match with today's requirements.

They have become less toxic for humans, they have become more specific, and they have become more powerful. For example, 40 years ago pesticides were applied in kilograms of litres of active ingredients per hectare. Modern pesticides only require grams or millilitres to achieve the same of better results.

Remember that pesticide residues remain only on the surface of produce. Rinsing with clear water will remove most of any trace residues.

PESTICIDE PROBLEMS

There are various problems which can occur when using synthetic chemicals to control pests. Some pests can build up resistance to pesticides. Pesticide resistance is the ability of an organism to tolerate a lethal level of a pesticide. This results in larger applications and a reduction in the effectiveness of the pesticide. Many of the chemicals used are not selective. They may also kill beneficial species, such as bees which pollinate crops.

Synthetic chemicals may end up hurting the plant the farmer is trying to preserve. Environmental pollution is another problem as pesticides may build up in the soil or run off in the water supply.

Pest resurgence may also occur. Pest resurgence occurs when a pest population level can establish itself after control measures against the population have been eliminated or reduced. In some countries other than Canada, the irrational use of pesticides and equipment has led to serious problems. In some crops, the costs of pesticide use are already higher than the benefits. While modern pesticides have reached the most remote parts of the world, the application equipment matches that which existed 40 years ago. This has resulted in large wastes of pesticides and unnecessary environmental contamination.

Farmers and application equipment operator's knowledge of pesticides and the correct method of application is usually deficient or non existent. In many cases they do not receive any training on the issue. Sometimes, the only specialists are representatives of pesticide companies. Of course it would be a poor business move for them to show farmers how to save money by using less of their product. Some farmers and spray equipment operators still believe that the more pesticides you use, the better. The lack of technical knowledge and awareness of farmers and manufacturers are the main reasons for the low quality of equipment found in some countries. There are also many cases of poor equipment maintenance. Many products which are considered to be highly or even extremely toxic and can lead to a considerable amount of poisoning are still used in some countries.

Pollutant

After a pesticide is applied not all of the pesticide remains where it is needed. When a pesticide leaves the target area, it is often considered to be an environmental pollutant. Movement of pesticides can occur in several ways. Drift, soil leaching, run-off, improper disposal and storage, and improper application are some of the major causes. Pesticide drift is a major cause of soil and air contamination. Drift occurs at the time of pesticide application when small spray particles are moved by air currents to nontargeted areas.

Most of these sources of contamination can be avoided with proper care and handling on the part of the farmer.

New Technology

Pressure for the use of pesticides in a manner that is safer for environment and humans has led to new approaches to pest control. Modern application equipment allows a fairly safe and efficient application of pesticides of all kinds. Legal requirements and farmer's demands have forced equipment manufacturers in Europe to offer only state of the art machinery. The equipment is designed to help prevent unnecessary contamination, accidents, losses and spills. The use of modern electronics has improved the accuracy of pesticide spraying. The Global Positioning Systems (GPS) for precise tracking has improved the safety of aerial pesticide spraying. GPS technology also allows patch spraying against certain weeds. If a weed is detected by the GPS system, the nozzle on the sprayer is opened only at that point. Therefore, lower amounts of pesticides are used. The entire field does not need to be dosed, which saves money for the farmer.

Alternatives

Non chemical alternatives for pest control have existed for quite some time. Some nonchemical alternatives include cultural practices such as crop rotation, choice of resistant plant varieties and use of biological products and agents. The concept of pest management, where synthetic pesticides are only applied as a last resort, is now considered common practice in professional agriculture. The development of plants with built-in pest resistance is a valuable alternative. This can be done by selecting plants with pest resistance, by crossbreeding a crop plant and a plant with a desirable characteristic, or by using plant biotechnology. However, some of these new developments have also encountered significant opposition.

ENVIRONMENTAL IMPACT OF PESTICIDES

The environmental impact of pesticides is often greater than what is intended by those who use them. Over 98 per cent of sprayed insecticides and 95 per cent of herbicides reach a destination other than their target species, including nontarget species, air, water, bottom sediments, and food. Though there can be benefits using pesticides, inappropriate use can counterproductively increase pest resistance and kill the natural enemies of pests. Many users are inadequately informed about potential short and long-term risks, and the necessary precautions in the correct application of such toxic chemicals are not always made. Pesticides can contaminate unintended land and water when they are sprayed aerially or allowed to run off fields, or when they escape from production sites and storage tanks or are inappropriately discarded. The amount of pesticide that migrates from the intended application area is influenced by the particular chemical's properties: its propensity for binding to soil, its vapor pressure, its water solubility, and its resistance to being broken down over time. Factors in the soil, such as its texture, its ability to retain water, and the amount of organic matter contained in it, also affect the amount of pesticide that will leave the area. Some pesticides contribute to global warming and the depletion of the ozone layer.

Air

Pesticides can contribute to air pollution. Pesticide drift occurs when pesticides suspended in the air as particles are carried by wind to other areas, potentially contaminating them. Pesticides that are applied to crops can volatilize and may be blown by winds into nearby areas, potentially posing a threat to wildlife. Weather conditions at the time of application as well as temperature and relative humidity change the spread of the pesticide in the air. As wind velocity increases so does the spray drift and exposure. Low relative humidity and high temperature result in more spray evaporating. The amount of inhalable pesticides in the outdoor environment is therefore often dependent on the season. Also, droplets of sprayed pesticides or particles from pesticides applied as dusts may travel on the wind to other areas, or pesticides may adhere to particles that blow in the wind, such as dust particles. Ground spraying produces less pesticide drift than aerial spraying does. Farmers can employ a buffer zone around their crop, consisting of empty land or non-crop plants such as evergreen trees to serve as windbreaks and absorb the pesticides, preventing drift into other areas. Such windbreaks are legally required in the Netherlands. Pesticides that are sprayed on to fields and used to fumigate soil can give off chemicals called volatile organic compounds, which can react with other chemicals and form a pollutant called tropospheric ozone. Pesticide use accounts for about 6 per cent of total tropospheric ozone levels.

Water

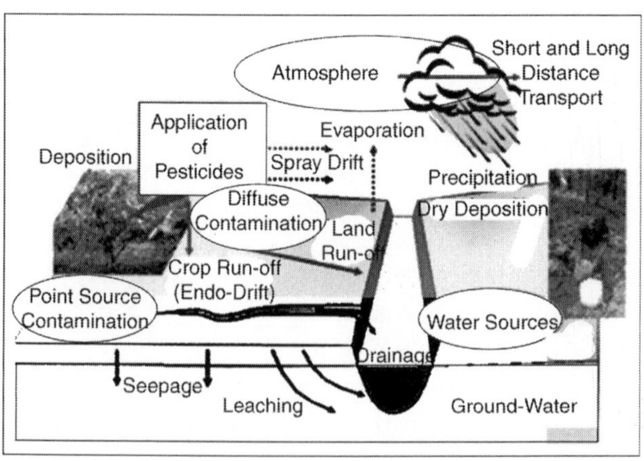

Fig. Pesticide Pathways

In the United States, pesticides were found to pollute every stream and over 90 per cent of wells sampled in a study by the US Geological Survey. Pesticide residues have also been found in rain and groundwater. Studies by the UK government showed that pesticide concentrations exceeded those allowable for drinking water in some samples of river water and groundwater.

Pesticide impacts on aquatic systems are often studied using a hydrology transport model to study movement and fate of chemicals in rivers and streams. As early as the 1970s quantitative analysis of pesticide runoff was conducted in order to predict amounts of pesticide that would reach surface waters.

There are four major routes through which pesticides reach the water: it may drift outside of the intended area when it is sprayed, it may percolate, or leach, through the soil, it may be carried to the water as runoff, or it may be spilled, for example accidentally or through neglect. They may also be carried to water by eroding soil. Factors that affect a pesticide's ability to contaminate water include its water solubility, the distance from an application site to a body of water, weather, soil type, presence of a growing crop, and the method used to apply the chemical.

Maximum limits of allowable concentrations for individual pesticides in public bodies of water are set by the Environmental Protection Agency in the US. Similarly, the government of the United Kingdom sets Environmental Quality Standards (EQS), or maximum allowable concentrations of some pesticides in bodies of water above which toxicity may occur. The European Union also regulates maximum concentrations of pesticides in water.

Soil

Many of the chemicals used in pesticides are persistent soil contaminants, whose impact may endure for decades and adversely affect soil conservation.

The use of pesticides decreases the general biodiversity in the soil. Not using the chemicals results in higher soil quality, with the additional effect that more organic matter in the soil allows for higher water retention. This helps increase yields for farms in drought years, when organic farms have had yields 20-40 per cent higher than their conventional counterparts. A smaller content of organic matter in the soil increases the amount of pesticide that will leave the area of application, because organic matter binds to and helps break down pesticides.

Degradation and sorption are both factors which influence the persistence of pesticides in soil. Depending on the chemical nature of the pesticide, such processes control directly the transportation from soil to water, and in turn to air and our food. Breaking down organic substances, degradation, involves interactions among microorganisms in the soil. Sorption affects bioaccumulation of pesticides which are dependant on organic matter in the soil. Weak organic acids have been shown to be weakly sorbed by soil, because of pH and mostly acidic structure. Sorbed chemicals have been shown to be less accessible to microorganisms. Aging mechanisms are poorly understood but as residence times in soil increase, pesticide residues become more resistant to degradation and extraction as they lose biological activity.

Effect on plants

Nitrogen fixation, which is required for the growth of higher plants, is hindered by pesticides in soil. The insecticides DDT, methyl parathion, and especially pentachlorophenol have been shown to interfere with legume-rhizobium chemical signaling. Reduction of this symbiotic chemical signaling results in reduced nitrogen fixation and thus reduced crop yields. Root nodule formation in these plants saves the world economy $10 billion in synthetic nitrogen fertilizer every year.

Pesticides can kill bees and are strongly implicated in pollinator decline, the loss of species that pollinate plants, including through the mechanism of Colony Collapse Disorder, in which worker bees from a beehive or Western honey bee colony abruptly disappear. Application of pesticides to crops that are in bloom can kill honeybees, which act as pollinators. The USDA and USFWS estimate that US farmers lose at least $200 million a year from reduced crop pollination because pesticides applied to fields eliminate about a fifth of honeybee colonies in the US and harm an additional 15 per cent.

On the other side, pesticides have some direct harmful effect on plant including poor root hair development, shoot yellowing and reduced plant growth .

Effect on animals

Pesticides inflict extremely widespread damage to biota, and many countries have acted to discourage pesticide usage through their Biodiversity Action Plans. Animals may be poisoned by pesticide residues that remain on food after spraying, for example when wild animals enter sprayed fields or nearby areas shortly after spraying.

Widespread application of pesticides can eliminate food sources that certain types of animals need, causing the animals to relocate, change their diet, or starve. Poisoning from pesticides can travel up the food chain; for example, birds can be harmed when they eat insects and worms that have consumed pesticides. Earthworms digest organic matter and increase nutrient content in the top layer of soil. They aid in protecting human health by ingesting decomposing litter and serving as bioindicators in soil activity while creating a richer environment. A number of studies have shown that pesticides have had harmful effects on growth and reproduction on earthworms, which are in turn consumed by terrestrial vertebrates such as birds and small mammals. Some pesticides can bioaccumulate, or build up to toxic levels in the bodies of organisms that consume them over time, a phenomenon that impacts species high on the food chain especially hard.

Birds

The Fish and Wildlife Service estimates that 72 million birds are killed by pesticides in the United States each year. Bald eagles are common examples of nontarget organisms that are impacted by pesticide use. Rachel Carson's landmark book *Silent Spring* dealt with the loss of bird species due to bioaccumulation of pesticides in their tissues. There is evidence that birds are continuing to be harmed by pesticide use. In the farmland of Britain, populations of ten different bird species declined by 10 million breeding individuals between 1979 and 1999, a phenomenon thought to have resulted from loss of plant and invertebrate species on which the birds feed. Throughout Europe, 116 species of birds are now threatened. Reductions in bird populations have been found to be associated with times and areas in which pesticides are used. DDE-induced egg shell thinning has especially affected European and North American bird populations. In another example, some types of fungicides used in peanut farming are only slightly toxic to birds and mammals, but may kill off earthworms, which can in turn reduce populations of the birds and mammals that feed on them. Some pesticides come in granular form, and birds and other wildlife may eat the granules, mistaking them for grains of food. A few granules of a pesticide is enough to kill a small bird.

The herbicide paraquat, when sprayed onto bird eggs, causes growth abnormalities in embryos and reduces the number of chicks that hatch successfully, but most herbicides do not directly cause much harm to birds. Herbicides may endanger bird populations by reducing their habitat.

Aquatic life

Fish and other aquatic biota may be harmed by pesticide-contaminated water. Pesticide surface runoff into rivers and streams can be highly lethal to aquatic life, sometimes killing all the fish in a particular stream.

Application of herbicides to bodies of water can cause fish kills when the dead plants rot and use up the water's oxygen, suffocating the fish. Some herbicides, such as copper sulfite, that are applied to water to kill plants are toxic to fish and other water animals at concentrations similar to those used to kill the plants. Repeated exposure to sublethal doses of some pesticides can cause physiological and behavioral changes in fish that reduce populations, such as abandonment of nests and broods, decreased immunity to disease, and increased failure to avoid predators. Application of herbicides to bodies of water can kill off plants on which fish depend for their habitat. Pesticides can accumulate in bodies of water to levels that kill off zooplankton, the main source of food for young fish. Pesticides can kill off the insects on which some fish feed, causing the fish to travel farther in search of food and exposing them to greater risk from predators. The faster a given pesticide breaks down in the environment, the less threat it poses to aquatic life. Insecticides are more toxic to aquatic life than herbicides and fungicides.

Amphibians

In the past several decades, decline in amphibian populations has been occurring all over the world, for unexplained reasons which are thought to be varied but of which pesticides may be a part. Mixtures of multiple pesticides appear to have a cumulative toxic effect on frogs. Tadpoles from ponds with multiple pesticides present in the water take longer to metamorphose into frogs and are smaller when they do, decreasing their ability to catch prey and avoid predators.

A Canadian study showed that exposing tadpoles to endosulfan, an organochloride pesticide at levels that are likely to be found in habitats near fields sprayed with the chemical kills the tadpoles and causes behavioral and growth abnormalities. The herbicide atrazine has been shown to turn male frogs into hermaphrodites, decreasing their ability to reproduce. Both reproductive and nonreproductive effects in aquatic reptiles and amphibians have been reported. Crocodiles, many turtle species, and some lizards lack sex distinct chromosomes until after fertilization during organogenesis, depending on temperature. Exposure to various PCBs in the embryonic stage of development in turtles has shown to cause a sex reversal. Across the United States and Canada disorders such as decreased hatching success, feminization, skin lesions, and other such developmental abnormalities have been reported.

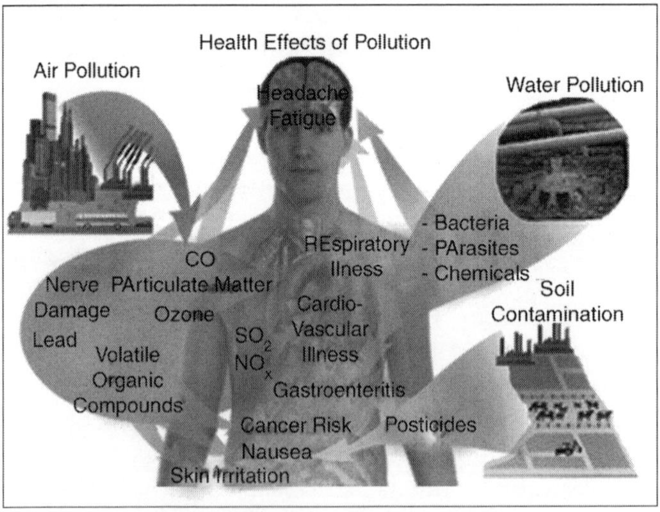

Fig. Pesticides are Implicated in a Range of Impacts on Human Health Due to Pollution

Humans

Pesticides can enter the human body through inhalation of aerosols, dust and vapor that contain pesticides; through oral exposure by consuming food and water; and through dermal exposure by direct contact of pesticides with skin. Pesticides are sprayed onto food, especially fruits and vegetables, they secrete into soils and groundwater which can end up in drinking water, and pesticide spray can drift and pollute the air.

The effects of pesticides on human health are more harmful based on the toxicity of the chemical and the length and magnitude of exposure. Farm workers and their families experience the greatest exposure to agricultural pesticides through direct contact with the chemicals. But every human contains a percentage of pesticides found in fat samples in their body. Children are more susceptible and sensitive to pesticides because they are still developing and have a weaker immune system than do adults. Children may be exposed due to their closer proximity to the floor and natural tendency to put contaminated objects in their mouth, and also because children tend to spend more time at home in a potentially contaminated environment. Hand to mouth contact will be dependent on the age of the child, much like lead exposure, typically from dust within the home. Children under the age of six months are more apt to experience exposure from breast milk and inhalation of small particles. Pesticides may be absorbed through dermal contact, ingestion, and inhalation. Pesticides tracked into the home from family members increase the risk of toxic pesticide exposure which is normally area specific. Also, toxic residue in food may contribute to a child's exposure to a certain pesticide. The chemicals can bioaccumulate in the body over time.

Exposure to pesticides can range from mild skin irritation to birth defects, tumors, genetic changes, blood and nerve disorders, endocrine disruption, and even coma or death. Developmental effects have been associated with pesticides. Recent increases in childhood cancers in throughout North America, such as leukemia, may be a result of genotoxic and nongenotoxic pesticides due to somatic cell mutations. Insecticides targeted to disrupt insects can have harmful effects on the nervous systems of mammals, due to basic similarities in system structure. Both chronic and acute alterations have been observed in those who are exposed. Pesticides can act in the promotion and proliferation of cancer while causing hormone imbalance. DDT and its breakdown product DDE, with levels still present in the environment, despite its ban, are known to disturb estrogenic activity and possibly lead to breast cancer. Exposure to pesticides, for example DDT, in fetal stages has been proven to alter male penis size in animals to that much smaller than average as well as develop undescended testicles. Exposure to pesticides may occur in postnatal early stages of development, in utero, and even if either parent was exposed before conception took place. Reproductive disruption has the potential to occur by chemical reactivity and through structural changes to a system.

Persistent Organic Pollutants

Persistent organic pollutants (POPs) are compounds that resist degradation and thus remain in the environment for years. Some pesticides, including aldrin, chlordane, DDT, dieldrin, endrin, heptachlor, hexachlorobenzene, mirex, and toxaphene, are considered POPs. POPs have the ability to volatilize and travel great distances through the atmosphere to become deposited in remote regions. The chemicals also have the ability to bioaccumulate and biomagnify, and can bioconcentrate (i.e. become more concentrated) up to 70,000 times their original concentrations. POPs may continue to poison non-target organisms in the environment and increase risk to humans by disruption in the endocrine, reproductive, and immune systems; cancer; neurobehavioral disorders, infertility and mutagenic effects, although very little is currently known about these chronic effects. Some POPs have been banned, while others continue to be used.

Pest Resistance

Pests may evolve to become resistant to pesticides. Many pests will initially be very susceptible to pesticides, but some with slight variations in their genetic makeup are resistant and therefore survive to reproduce. Through natural selection, the pests may eventually become very resistant to the pesticide.

Pest resistance to a pesticide is commonly managed through pesticide rotation, which involves alternating among pesticide classes with different modes of action to delay the onset of or mitigate existing pest resistance.

Pest Rebound and Secondary Pest Outbreaks

Non-target organisms, organisms that the pesticides are not intended to kill, can be severely impacted by use of the chemicals. In some cases, where a pest insect has some controls from a beneficial predator or parasite, an insecticide application can kill both pest and beneficial populations. A study comparing biological pest control and use of pyrethroid insecticide for diamondback moths, a major cabbage family insect pest, showed that the insecticide application created a rebounded pest population due to loss of insect predators, whereas the biocontrol did not show the same effect. Likewise, pesticides sprayed in an effort to control adult mosquitoes, may temporarily depress mosquito populations, however they may result in a larger population in the long run by damaging the natural controlling factors. This phenomenon, wherein the population of a pest species rebounds to equal or greater numbers than it had before pesticide use, is called pest resurgence and can be linked to elimination of predators and other natural enemies of the pest.

Loss of predator species can also lead to a related phenomenon called secondary pest outbreaks, an increase in problems from species which were not originally very damaging pests due to loss of their predators or parasites. An estimated third of the 300 most damaging insects in the US were originally secondary pests and only became a major problem after the use of pesticides. In both pest resurgence and secondary pest outbreaks, the natural enemies have been found to be more susceptible to the pesticides than the pests themselves, in some cases causing the pest population to be higher than it was before the use of pesticide.

Eliminating Pesticides

Many alternatives are available to reduce the effects pesticides have on the environment. There are a variety of alternative pesticides such as manually removing weeds and pests from plants, applying heat, covering weeds with plastic, and placing traps and lures to catch or move pests. Pests can be prevented by removing pest breeding sites, maintaining healthy soils which breed healthy plants that are resistant to pests, planting native species that are naturally more resistant to native pests, and use biocontrol agents such as birds and other pest eating organisms.

There is much debate over alternative methods of controlling pesticides. In many instances, biological controls such as resistant plant varieties and the use of pheromones, have been successful and at times resolving a pest problem permanently. Integrated Pest Management (IPM) has also been one practice in pest management which establishes chemical use on a need basis only. IPM is a more socially accepted practice and causes less harm to the health of humans and the environment. The focus is on the pest culture, life cycle, and role in the environment. Biotechnology can also be an innovative way to control pests. Technological advances, such as genetically modified (GM) plants, may have a stronger resistance to pests and could eliminate pesticides use in the future.

CHEMICAL CONTROL OF PESTICIDES

Pesticides are regarded by many as controls of last resort because they add considerably to the cost of production and their misuse creates high-profile environmental problems. Nevertheless, in well-planned, integrated pest management programmes, pesticides have a valuable role if used judiciously.

A major problem with many pesticides is the rapid development of pathogen races or insect and mite populations that are resistant to them. When that happens, their continued use is pointless; not only do they have little or no efficacy, but they can aggravate the situation by killing other organisms that contribute to the natural biological control of pathogens and other pests. Blanket, season-long pesticide programmes used as insurance in anticipation of pest outbreaks are seldom justified if other precautions have been taken.

The application of crop protection chemicals to vegetable crops or to the soil is a widely used method of controlling diseases, insects, weeds and many other pests. Because of the environmental impact of the use of chemical pesticides and because of their toxicity to humans, management procedures to minimize their use and their effects on non-target species are being developed and implemented. Accordingly, applications of effective pesticidal chemicals should be made in response to indications that populations of the pest have developed or will develop to economically significant densities. The threshold at which action is taken is dependent on the type of damage inflicted on the crop, especially the marketed portion of the crop. Populations of pests on crop plants are monitored to determine the timing and necessity for applications. This procedure of pest management minimizes the amount of chemical applied throughout the season.

Unfortunately, many growers prefer to apply pesticides indiscriminately to crops according to a predetermined schedule as insurance against damage by pests. This practice increases the detrimental environmental impact of adding pesticide chemicals to the field habitat and increases the cost of crop production. Also, the application of some chemicals causes stress to crop plants and may make them more susceptible to disease and other pests. On the other hand, scheduled applications of pesticides may be necessary to prevent development of economically damaging populations of pests, such as root maggots and other pests within roots or stems, that cannot readily be killed when crop damage is occurring, or pests for which populations cannot be predicted with acceptable accuracy.

Environmental impact: The unnecessary use of pesticides is discouraged because of their direct and indirect effects on the field habitat and environment, on the consumer and the grower, and on the cost and efficiency of production. Most insecticidal and acaricidal chemicals kill or are harmful to several species of non-target arthropods, including parasitic and predaceous species that contribute to regulation of primary or potential pests. There are cases where additional applications of pesticides were required to control pests whose populations increased to a damaging level following application of a pesticide to control a different species. Applications of chemicals may affect populations of birds and other animals that feed on pest species, and residues of some chemicals in soil and on plant debris may affect microbial activity in soil, thus influencing the availability of plant nutrients. In addition, there is concern for the possible harmful effects of residues of chemical pesticides in food and for the effects on growers of excessive exposure. To chemicals. Also, the tendency for pest species to become resistant or tolerant to a pesticide is enhanced by repeated use of the chemical, necessitating the use of higher concentrations or the use of other chemicals, often resulting in greater crop loss and increased cost of production.

MANAGEMENT BY EXCLUSION AND REGULATION

Federal legislation administered by Agriculture and Agri-Food Canada and complemented by provincial legislation is the means by which certain diseases and pests are regulated in Canada. Such regulations are aimed at preventing the introduction of new diseases and pests and their movement within Canada. The Canada Seeds Act and the Plant Protection Act provide for field surveys; inspection of produce, premises, containers and packages; destruction or treatment; restriction on movement; and other safeguards and requirements as specified in regulations under these Acts. Treatment of infested commodities, short of destruction, consists of fumigation, cleaning in a manner that is satisfactory to an inspector, or diversion to an approved industrial process. There are substantial reasons for concern about pests that may be associated with vegetable commodities at the time of harvest.

For instance, pests may be disseminated to other production regions within the country, a possibility that is greatly enhanced by modern, rapid transportation facilities. Pests also have resting stages that may survive cleaning and packaging operations and storage. The harvested, unprocessed commodity itself may serve as a pathway for the spread of pests. Apart from cold storage, which may affect the survival of some pests, Canada does not yet have the necessary facilities for bulk treatment and disinfestation of commercially produced quantities of vegetable commodities. Research into such alternative pest control technologies has only just begun.

Further information about the current status of federally regulated pests can be obtained from Agriculture and Agri-Food Canada.

Potato Viruses

- Andean potato latent virus
- Andean potato mottle virus
- Arracacha virus B
- Tobacco ringspot virus, potato calico strain
- Potato deforming mosaic virus
- Potato mop top virus
- Potato virus T
- Potato virus V
- Potato virus X, resistance breaking strain.

These and other viruses that occur in other countries are of concern to Canada because of their potential effects on crop yield and impact on trade. Their importation into Canada poses a threat to the potato industry in Canada, and thus they are cause for regulatory concern. To protect Canadian producers from potentially harmful viruses, the importation of potatoes from all countries other than the USA is prohibited. However, germplasm for research or commercial evaluation can be imported through Agriculture Canada's post-entry quarantine programme.

The Columbia root-knot nematode has not been reported in Canada, but it is widely distributed in the western United States. It was described as a distinct species in 1980 after years of confusion with the northern root-knot nematode, which is widespread in Canada. The Columbia root-knot nematode is a pathogen of bean, carrot, corn, pea and potato. It also affects canola, sugar beet and cereals, such as wheat, barley and oat. On potato, it induces galls on the surface of the tubers, greatly downgrading them. When tubers are sliced, the body of the female nematode can be seen as a tiny, water-soaked dot, and masses of eggs appear as small brown spots.

Potato-rot Nematode Ditylenchus Destructor Thorne

The potato-rot nematode is a widespread and serious pest of potato in Europe; it also occurs in many Asian countries and has been reported from New Zealand and Peru. In South Africa it is an important pest of peanut. In the United States, it has been found in Arizona, California, Hawaii, Idaho, Indiana, New Jersey, Oregon, Washington and Wisconsin, and it is subject to quarantine regulations in some of those states. In Canada, the nematode was detected in potato in 1945 in three small areas of Prince Edward Island, and in 1952 in a small potato planting on Vancouver Island, British Columbia. Following eradication procedures and provincial legislation, the infested fields in P.E.I. have been kept out of potato production, and the nematode has not been detected in surveys conducted since the early 1960s.

The potato-rot nematode also attacks bulbous iris and over 80 other hosts, including such vegetables as beet, carrot, celery, crucifers, cucurbits, eggplant, onion, pepper, rhubarb and tomato. Information on host range, diagnostic techniques, feeding habits, ecology and persistence of *D. destructor* in the field is incomplete; most infestations are detected only when found in mature potato tubers.

The potato-rot nematode feeds chiefly on fungi and can survive in weed hosts and in fallow soil for several years, but it seldom survives in rotted tubers in storage or in the field. Infestations are spread mainly by seed tubers. The nematode feeds first on the roots and later invades the tubers, in which it feeds just under the skin.

Initial symptoms include small (0.3 mm diameter) pits surrounded by snow-white rings that are visible when the tuber is peeled. Continued feeding results in a depressed area in the skin, which becomes necrotic and dries out, resulting in characteristic irregular, triangular cracks. Severely affected tubers have large, sunken areas covered by dried-out skin (16.37). The cracks in the skin provide ready access to bacteria and fungi, and affected tubers usually rot in storage or in the soil.

Tomato Pinworm Keiferia Lycopersicella (Walsingham)

The tomato pinworm (family Gelechiidae) occurs in Mexico, the southern United States, the Caribbean, and Hawaii. Larvae of this moth feed in developing tomato fruits and may be encountered on post-harvest tomatoes as an imported pest. The most recent importation into Canada occurred during 1988–89, when larvae were found in or on tomatoes originating from:

- *California:* This moth was unknown in Canada prior to 1946, when larvae were confirmed in field tomato crops and in greenhouses in southwestern Ontario; however, these infestations did not survive. There since have been two isolated infestations in Canada: in 1970 in a greenhouse on Vancouver Island, and in 1975 in a greenhouse and surrounding home gardens at Kamloops,
- *British Columbia:* Eradication at both locations was accomplished by exposure to the elements and by planting cucumber, which is a non-host crop. These occurrences probably had their origin in post-harvest tomatoes that then were being imported during the off-season from Mexico and the southern and western United States. These experiences show that the tomato pinworm cannot survive the winter outdoors.
- *Canada:* Temporary greenhouse infestations can be avoided by destroying used shipping containers and by screening greenhouse openings, and they can be managed by including cucumber or other non-solanaceous plants in the greenhouse cropping cycle.

BIOLOGICAL CONTROL USES NATURAL AGENTS

BIOLOGICAL CONTROL

Biological control uses natural agents such as predators, parasites, and pathogens to control pests. The natural enemy of an insect can be introduced to the crop.

For example, the aphid is a pest which eats many crops. Lady bugs eat aphids. Animals such as this which control pests and help a crop are called beneficial species. There can be some problems with this method. It might take more time than is needed for the farmer to save his crop, it may not be affordable, and it is sometimes hard to predict if the beneficial species will have other adverse effects on the environment. Biological control requires a thorough understanding of the biology and ecology of the beneficial organism, as well as the pest.

Organic Farming

Organic farming involves growing crops without applying synthetic fertilizers, herbicides, or pesticides. Only natural products, such as composted manure, are used to improve soil quality. Many natural products are used to control pest populations. For example, insecticidal soaps and oils are effective at controlling many different insects. These soaps and oils are non toxic to humans and pets. Botanical pesticides are derived from raw plant materials like flowers, roots, and stems. They are also safe to humans. Both types of chemicals can damage plants and kill beneficial organisms if care is not taken in their use. Here are two recipes for making these chemicals:

Soap and oil	Garlic Oil
1 Tbsp Liquid Dish Soap	1/2 Cup Minced Garlic
1 Cup Oil (Peanut, Corn, Soybean, or Sunflower)	2 Tsp Mineral Oil
1 Cup Water	1200 ml Water
	1 Tsp Liquid Dish Soap

Organic agriculture appears to be gaining acceptance on Prince Edward Island. There are more than 30 certified organic growers on the Island. More than 1,000 acres of land are currently being used to grow crops such as apples, blueberries, vegetables, herbs, soybeans, potatoes, hay, buckwheat and grain.

NATURAL FOOD SUBSTANCES

Many people are alarmed by the addition of man-made products, including small amounts of regulated pesticide residues, to their food. These people may be alarmed to find out that many substances which occur naturally in food can be just as harmful. Food-borne disease easily affects the most people. Usually this is due to the way food is handled or stored. Over consumption, poor food choices and less than desirable intake of many essential ingredients is also a major problem. Natural toxins in foods can be a problem when a person's diet is poor. Hazards from pesticide residues or food additives are only a minor problem when it comes to harmful foods. Most people would view these hazards virtually in reverse order of importance. Chemicals used as pesticides are popularly considered the key threat to food safety by many. Lack of education on the subject and media propaganda are responsible for this.

Carcinogens

The residues of a few pesticides used on food crops can pose potential health problems by acting as carcinogens. A carcinogen is a material capable of causing cancer. Some recent studies seem to indicate that many natural compounds in food may pose greater cancer risks than man-made carcinogens. People should rest assured that the actual cancer risk from both natural and man-made compounds is minutely small. All of the food which we eat is tested very carefully. Simply eating a balanced diet that includes most fruit and vegetables is actually a good way to fend off cancer.

Pesticides in Canada

Canada's pesticide regulatory system is one of the most stringent in the world. The Food and Drugs Act in this country limits pesticide residues in foods to no more than 0.1 parts per million. Each year, the federal government tests thousands of fruit and vegetable samples to ensure maximum residue limits are not exceeded. Any imported produce must meet the same standards.

Responsibility for pesticide regulations is administered by Health Canada. Before being registered for use, all pesticides are approved by the Pest Management Regulatory Agency under Health Canada. Pesticide use in Canada is lower than most countries. Canada uses less than one twentieth of what is used in the Netherlands, a fifth of what is used in France and less than half of what is used in the United States.

Local Concerns

There is a growing concern about the known and unknown consequences of pesticides on environment and human health. This has led to a strong public pressure to reduce their use. Some farmers on Prince Edward Island have already encountered this pressure. In 1996, some residents who live along the Mason Road in Stratford complained that drift from the spraying of nearby potato fields was making them ill. A study by the Provincial Health Officer found no relation between pesticide sprays and the symptoms experienced by these residents. During the summer of 1997, the Town of Stratford presented its official plan to the provincial Attorney General. It called for a ban on "intensive agriculture" practices in the town's residential zone. This would put an end to potato farming in Stratford. After consulting with the town and agriculture groups, the Attorney General signed the plan in July. As of August 1997, the term "intensive agriculture" was still in the process of being clearly defined. Many farmers around the Island are concerned about the precedent that Stratford is attempting to set. The general sentiment is that if the ban is allowed in Stratford, what will stop other communities from following the same example.

Education is the Key

The lack of knowledge at all levels has been identified as the main reason for deficiencies in pesticide application practice. Because of the Pesticide Act on Prince Edward Island, farm owners and managers are required to be licensed as pesticide applicators. Practical training of farmers and equipment operators has been introduced. Continuing education will provide the technical knowledge necessary for a safe and efficient application by the industry. The public also needs to become educated about the benefits and drawbacks of pesticide use. This would allow them to form their own opinions instead of being influenced by what they hear on television or read in the news.

PESTICIDE RESISTANCE

Pesticide resistance is the adaptation of pest population targeted by a pesticide resulting in decreased susceptibility to that chemical. In other words, pests develop a resistance to a chemical through natural selection: the most resistant organisms are the ones to survive and pass on their genetic traits to their offspring. Manufacturers of pesticides tend to prefer a definition that is dependent on failure of a product in a real situation, sometimes called field resistance. For example, the Insecticide Resistance Action Committee (IRAC) definition of insecticide resistance is 'a heritable change in the sensitivity of a pest population that is reflected in the repeated failure of a product to achieve the expected level of control when used according to the label recommendation for that pest species'.

Pesticide resistance is increasing in occurrence. In the 1940s, farmers in the USA lost 7 per cent of their crops to pests, while since the 1980s, the percentage lost has increased to 13, even though more pesticides are being used. Over 500 species of pests have developed a resistance to a pesticide. Other sources estimate the number to be around 1000 species since 1945.

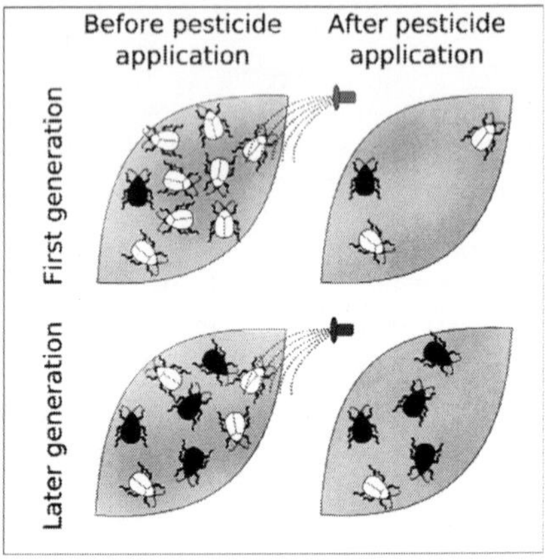

Fig. Pesticide Application can Artificially Select for Resistant Pests.

In this diagram, the first generation happens to have an insect with a heightened resistance to a pesticide. After pesticide application, its descendants represent a larger proportion of the population because sensitive pests have been selectively killed. After repeated applications, resistant pests may comprise the majority of the population.

Factors

Propensity of pest populations to develop resistance is probably caused by a number of factors. First, pest species are usually capable of producing large number of offspring. This increases the probability of random mutations and ensures the rapid build-up in numbers of resistant mutants once such mutations have occurred. Secondly, pest species have been exposed to natural toxins for a long time before the onset of human civilization. For example, many plants produce phytotoxins to protect them from herbivores. As a result, coevolution of herbivores and their host plants required development of the physiological capability to detoxify or tolerate poisons. Thirdly, humans often rely almost exclusively on insecticides for pest control. This increases selection pressure towards resistance. Pesticides that fail to break down quickly and remain in the area contribute to selection for resistant organisms even after they are no longer being applied.

In response to pesticide resistance, pest managers may resort to increased use of pesticides, exacerbating the problem. In addition, when pesticides are toxic toward species that feed on or compete with pests, the pest population will likely expand further, requiring more pesticides. This is sometimes referred to as *pesticide trap*, or a *pesticide treadmill*, since farmers are continually paying more for less benefit.

Insect predators and parasites which live on other insects generally have smaller populations and are therefore much less likely to develop resistance than are the primary targets of the pesticides, such as mosquitoes and those that feed on plants. This can compound the pest problem because these species normally keep pest populations in check. But resistant predators of pest species can be bred in laboratories, which can help keep pest populations down.

The fewer sources of food a pest has the more likely it is to develop resistance, because it is exposed to higher concentrations of pesticides and has less opportunity to breed with populations that have not been exposed. Other factors in the speed with which a species develops resistance are generation time and fecundity (shorter generations and more offspring lead to resistance more quickly).

Examples

Resistance has developed in a variety of different pest species: Resistance to insecticides was first documented by A. L. Melander in 1914 when scale insects demonstrated resistance to an inorganic insecticide. Between 1914 and 1946, 11 additional cases of resistance to inorganic insecticides were recorded. The development of organic insecticides, such as DDT, gave hope that insecticide resistance was an issue of the past. Unfortunately, by 1947 housefly resistance to DDT was documented. With the introduction of every new insecticide class – cyclodienes, carbamates, formamidines, organophosphates, pyrethroids, even *Bacillus thuringiensis* – cases of resistance surfaced within two to 20 years.

- In the US, studies have shown that fruit flies that infest orange groves were becoming resistant to malathion, a pesticide used to kill them.
- In Hawaii and Japan, the diamondback moth developed a resistance to *Bacillus thuringiensis* about three years after it began to be used heavily.
- In England, rats in certain areas have developed such a strong resistance to rat poison that they can consume up to five times as much of it as normal rats without dying.
- DDT is no longer effective in preventing malaria in some places, a fact which contributed to a resurgence of the disease.

- In the southern United States, the weed Amaranthus palmeri, which interferes with production of cotton, has developed widespread resistance to the herbicide Roundup.
- Colorado potato beetle has developed resistance to 52 different compounds belonging to all major insecticide classes. Resistance levels vary greatly among different populations and between beetle life stages, but in some cases can be very high (up to 2,000-fold).

Although resistance development is usually discussed as a result of pesticide use, it is important to keep in mind that pest populations can also adapt to non-chemical methods of control. For example, the northern corn rootworm (*Diabrotica barberi*) became adapted to a corn-soybean crop rotation by spending the year when field is planted to soybeans in a diapause.

Multiple and Cross-resistance

Multiple resistance is the phenomenon in which a pest is resistant to more than one class of pesticides. This can happen if one pesticide is used until pests display a resistance and then another is used until they are resistant to that one, and so on. Cross resistance, a related phenomenon, occurs when the genetic mutation that made the pest resistant to one pesticide also makes it resistant to other pesticides, especially ones with similar mechanisms of action.

Physiology and Behavior

Frequently a pest becomes resistant to a pesticide because it develops physiological changes that protect it from the chemical. In some cases, a pest may gain an increased number of copies of a gene, allowing it to produce more of a protective enzyme that breaks down the pesticide into less toxic chemicals. Such enzymes include esterases, glutathione transferases, and mixed microsomal oxidases. Alternatively, the number of biochemical receptors for the chemicals may be reduced in the pest, or the receptor may be altered, reducing the pest's sensitivity to the compound. Behavioral resistance has also been described for some chemicals; for example, some Anopheles mosquitoes developed a preference for resting outside that prevented them from coming in contact with pesticide sprayed on interior walls. Still other mechanisms include increased rates of excretion of toxic molecules, their sequestration and storage inside of the insect body away from vulnerable tissues and organs, and decreased toxin penetration through the insect body wall.

Often, mutation in a single gene only leads to the development of a resistant organism. In other cases, multiple genes are involved. Resistant genes are usually autosomal. This means that they are located on autosomes (as opposed to sex chromosomes). As a result, resistance is inherited similarly in males and females. Also, resistance is usually inherited as an incompletely dominant trait. When a resistant and a susceptible individual mate with each other, their progeny has an intermediate level of resistance (more resistant than the susceptible parent, but not as resistant as the resistant parent).

Adaptation to pesticides usually decreases relative fitness of organisms in the absence of pesticides. Resistant individuals often have reduced reproductive output, life expectancy, mobility, etc. Therefore, relatively few of them persist in a population that is not exposed to a particular insecticide to which they have developed resistance.

Blowfly maggots produce an enzyme that confers resistance to organochloride insecticides. Scientists have researched ways to use this enzyme to break down pesticides in the environment, which would detoxify them and prevent harmful environmental effects. Later they discovered a similar enzyme produced by soil bacteria that also breaks down organochloride insecticides but which works faster and remains stable in a variety of conditions. The product, called Landguard is used in Australia to decontaminate spray equipment, soil and water after pesticide spraying and spills.

Management

Pest resistance to a pesticide can be managed by reducing selection pressure by this pesticide on the pest population. In other words, the situation when all the pests except the most resistant ones are killed by a given chemical should be avoided. This can be achieved by avoiding unnecessary pesticide applications, using non-chemical control techniques, and leaving untreated refuges where susceptible pests can survive. Adopting the integrated pest management (IPM) approach usually helps with resistance management.

When pesticides are the sole or predominant method of pest control, resistance is commonly managed through pesticide rotation. This involves alternating among pesticide classes with different modes of action to delay the onset of or mitigate existing pest resistance. Different pesticide classes may have different effects on a pest. The U.S. Environmental Protection Agency (EPA or USEPA) designates different classes of fungicides, herbicides and insecticides. Pesticide manufacturers may, on product labeling, require that no more than a specified number of consecutive applications of a pesticide class be made before alternating to a different pesticide class. This manufacturer requirement is intended to extend the useful life of a product.

Tankmixing pesticides is the combination of two or more pesticides with different modes of action in order to improve individual pesticide application results and delay the onset of or mitigate existing pest resistance.

3

Nonpesticide Methods for Managing Crop Diseases

The problems highlighted above can be greatly reduced by using an integrated approach based on some of the nonpesticide methods of disease control that will be outlined in the following section of this paper. This will start with a discussion of methods used to prevent the entry of a pathogen to a new area or attempts to quickly eradicate such an incursion. Both approaches highlight the cost benefits achieved in preventing disease establishment and they should be the cornerstone of nonpesticide approaches to disease management in all countries of the region. This is followed by a description of certification schemes to provide pathogenfree plants to farmers and an in-depth examination of various cultural and physical practices and biological control methods that may reduce disease levels before discussing the importance of host plant resistance for achieving durable and sustainable disease control. The section concludes with an examination of the role of biotechnology and its application to disease control. It is stressed that no one method or approach to disease management holds the key to future success rather that solutions will be location-specific and will depend on a basket of options in order to achieve sustainable disease management. The problems and issues specific to all methods are discussed with possible strategies for future improvement. The paper highlights a number of case studies that highlight the impact of some nonpesticide approaches to disease management.

Regulatory Methods for Managing Crop Diseases – the Role of Quarantine and Eradication

Regulatory methods of disease management have as their objective measures that achieve exclusion of the pathogen from its host or a specific geographic area. This could be an entire country or a region within a country. These methods are enforced by regulations or acts of quarantine. Preventing the pathogen from coming into contact with its host will ensure that disease does not develop. Such an approach should be the foundation for the effective control of plant disease. It is the most cost-effective method for disease control and benefits all farmers in the region. The exclusion of banana bunchy top virus (BBTV) from PNG or TLB from Fiji saves both countries millions of dollars yearly. Unfortunately, exclusion is becoming increasingly difficult with the escalation in tourism, travel and commerce. There are growing demands on an effective quarantine service to support this objective. However, at the same time detection methods are becoming more reliable as sensitive diagnostics become available.

The island countries of the Pacific have a comparative advantage when it comes to exclusion. The island nature and isolation of many of the countries provides natural barriers to the movement of plant pathogens and make it easier for these countries to police the movement of pathogens and plants compared to countries in Asia that share a common border. Consequently, many of the island nations of the Pacific such as Fiji, Samoa, Tonga grow crops that are still free from major diseases. However, this comparative advantage is only as good as the enforcement and there are a number of constraints that face quarantine services in the Asian-Pacific region in the face of increasing trade and commerce. In recent years there have been a few notable cases of exotic diseases introduced to certain countries in the region with devastating impact.

TLB is a good example. Most recently, TLB spread to Samoa and American Samoa in 1993, most likely from Hawaii, devastating production. Within a matter of months the disease had caused millions of dollars in lost production. Pathogens introduced into an area in which they did not previously exist often have a dramatic impact. This is because the plants that the pathogens attack have not had the opportunity to develop any significant level of resistance in the absence of the pathogen. The nearby islands of Fiji, Cook Islands, Tonga, Wallis and Futuna and Niue have important taro industries and for the time being remain free from the disease but constant vigilance is required by quarantine services.

At the western side of PNG, along the border of West Papua, villagers are dealing with many plant diseases that seriously constrain smallholder production across the country. In addition, they are facing the added difficulties of having to deal with new diseases and threats to food security that are coming over the border from West Papua and further afield. Of great importance are blood disease of banana and citrus huanglongbing (greening) disease.

One other banana disease, fusarium wilt, has recently been introduced to PNG and could be catastrophic. Every effort needs to be made to prevent incursions of blood disease and citrus huanglongbing to PNG and the further spread of fusarium wilt to other parts of PNG. Likewise, devastating taro virus diseases such as *alomae* and *bobone,* currently confined to Melanesian countries pose a major quarantine threat to Indonesia and other Asian countries where this crop is important.

THE BENEFIT OF ERADICATION

Despite all the noble efforts of our quarantine services there are going to times when new pathogens are introduced. Even at this stage an effective emergency response by quarantine and plant protection personnel can reduce the chances of a pathogen establishing in a country. The cost-benefit analysis of this approach highlights the advantages. In 1984 a survey detected citrus canker on Thursday Island. This disease has the potential to destroy the citrus industry in the moist coastal production areas of Australia. Eradication of the disease from the island involved the destruction of several trees and detailed monitoring of the remaining trees for four years. Thursday Island was declared free of citrus canker in 1988 at a cost of US$20,000. Similar eradication programs have been successful for black sigatoka.

Crop Certification and the Use of Pathogen-Free Planting Material

A host plant will grow better and produce a higher yield if the planting material used is pathogen-free. For this reason every effort should be made to ensure that farmers get access to such planting material. In developed countries certification schemes are well developed for potatoes and certain fruit species and the provision of certified stock has led to significant improvements in general health and production. Similar certification schemes or programs to produce pathogen-free planting material exist in developing countries. In the Asia-Pacific region, Taiwan has an extensive range of schemes covering staple crops such as rice and coarse grains such as corn, soybean and peanut.

Other propagation and inspection systems of virus-free seedlings are now in place for potato, strawberry, sweet potato, asparagus, passion fruit, banana, citrus, sugarcane, anthurium, oncidium and bamboo. Most other countries in Asia have seed certification schemes, backed up by appropriate seed laws, for one or more crops, which have a degree of seed health testing to control the possibility of seed transmitted pathogens. In the Pacific region the formal seed supply network is poorly developed and there are no seed laws or quality standards. Some farmers have access to certified seed through agricultural merchants but generally most use farmer-saved seed.

MANAGING THE CROP

SANITATION

After genetic resistance, prophylaxis is by far the most effective and cheapest way of escaping major disease epidemics and pest infestations. It reduces the need for multiple applications of pesticides (which stress the crop), the risks of pesticide resistance, and pesticide contamination of the produce, the operator and the environment. Physical screening against immigrant pests has already been discussed, which, coupled with aggressive control of insects in the environs of the greenhouse and in adjacent weeds and field crops, is very effective prophylaxis against both direct damage and insect-transmitted diseases. Some growers rely on old crop prunings to perpetuate populations of biocontrol insects.

This is not a good practice because they constitute a reservoir of pathogens and non-parasitized pests. New introductions of biocontrol insects are a better practice. Reducing inoculum is also important in early crop management, with such tactics as quarantine, seed disinfestation, the use of healthy mother plants for cuttings, micropropagation, removing and properly disposing of all previous crop debris, pasteurizing or solarizing soil and soilless media, and disinfesting the greenhouse structure, benches, trays, stakes and other materials. Disinfestants include formaldehyde (as formalin) and hypochlorites but both materials are hazardous to humans and residues are phytotoxic. A persulphate oxidising agent, however, destroys viruses and micro-organisms without such side effects.

Crop Scheduling

Seeding, pricking-out and sticking cuttings should all be done in a greenhouse separate from the main production areas, and on mesh or slatted benches allowing through-the-bench ventilation. The benches should be well above the level of soilsplash and there should be no overhead pots from which contaminated soil and drainagewater fall.

Where there is risk of diseases more destructive in cool soils, for example, Fusarium crown and root rot and corky root rot of tomatoes, transplanting should be delayed until the root zone has warmed up, and insulating mulch materials put down later.

Where two or more monocrops are grown each year, overlapping of transplant production and marketable crop production means that pest and pathogen populations are perpetuated unless special care is taken to keep the young and cropping plants entirely separate. There is further risk if adjacent field crops constitute a reservoir of pathogens and pests.

Spacing

Close horizontal and vertical spacing of plants both on the bench and in the ground bed invites rapid plant-to-plant spread of walking insects, and of pathogens as diverse as *Pythium* spp., tomato mosaic virus, *Clavibacter michiganensis* Davis ssp. *michiganensis* Davis, the downy mildews and *Botrytis cinerea* Pers.: Fr. The agents of virus spread are mainly water and soil splash, insects, and workers handling plants with contaminated tools and fingers. Since air movement is restricted in dense plantings, the movement of airborne propagules 'is restricted, giving patchy distribution of diseases and insects. Moreover, close spacing results in undue inteiplant competition for water, nutrients, light and and undue damage by workers.

The Growing Medium

Growing media cover a wide spectrum of substrates: soil and soil-mix composts, organic materials such as sawdust and coconut fibre, inorganic materials such as rockwool and synthetic foams and aggregates, and the nutrient film technique (NFT). Soilborne diseases are no less prevalent in soilless substrates than in soil.

All substrates must be substantially free of insects and pathogens at planting and must be kept so throughout the life of the crop, thus demanding a high standard of hygiene. Soils are usually heavily amended with peat, farmyard manure, straw or crop residues. Ploughing or rotovating the soil should be done in order to comminute plant root debris and other organic matter, and so expose pathogen propagules to natural biological control. Getting the soil into good tilth with optimum temperature, water content and aeration promotes this microbial activity.

Soils also harbour several insects, such as pupae of leaf miners and thrips, as well as fungus gnat and shore fly larvae, both of which vector *Pythium* and *Fusarium* spp. Their populations, as well as populations of predatory microarthropods, are determined by soil organic matter, soil type and pore size. Populations of omnivorous collembola and non-cryptostigmatic mites, for example, are enhanced by the organic matter usually plentifully added to greenhouse soils. Fungal parasites of insects and nematodes are also encouraged in soils of good tilth.

The root-knot nematode *Meloidogyne incognita* Chitwood, however, survives at 1–2 m, well below soil disturbance levels. Most substrates can be fumigated or heat-sterilized but pasteurization to about 70 °C or serialization to about 40–55 °C is preferred over total steam sterilization to 100 °C because it preserves thermophilic biocontrol organisms. The whole greenhouse can be closed in sunny conditions for solarization of both substrate and superstructure.

High temperature and vapour pressure deficit in closed greenhouses can kill the western flower thrips (*F. occidentals* but unfortunately also its predator *Neoseiulus (=Amblyseius) cucumeris* (Oudemans). As with the original ideas that soilless cultivation would eliminate soilbome pathogens, crops in rockwool or other inert substrate, or in NFT are no less free of soilbome arthropods. Fungus gnats, leafminers and thrips are numerous in rockwool and shore flies are always present in pools of water on plastic sheets. Even if soil is covered with plastic sheet, there are always gaps around stems, and tears and displacement of the cover readily permit insect access.

Nutrition

Deficiencies and excesses of macro- and micronutrients, and imbalances in relative amounts of fertilizers can predispose plants to most diseases. In addition, fertilizers that increase foliage density at the expense of flowers and fruit not only reduce yield but tend to lower the vapour pressure deficit (VPD) in the boundary layer by restricting transpiration and wind-assisted evaporation, and consequently increase the risks of infection. High nitrogen rates in fertilizers generally increase foliage density and softness, with increasing susceptibility to leaf and flower pathogens.

For example, Hobbs and Waters found a quadratic increase in grey mould (*B. cinerea*) in chrysanthemum flowers (*Dendranthema grandiflora* Tzvelev) with nitrogen supplied with 1.5, 3.8 and 60 g/m^2. Nitrate nitrogen combined with liming gives excellent control of Fusarium wilt of several crops. Because of its role in the. integrity of cell walls, calcium imparts resistance if balanced with potassium in a high ratio. A low Ca:K ratio permits susceptibility to *B. cinerea* in tomato.

The K:N ratio is important in the susceptibility of tomato stems to the soft rot bacterium *Erwinia carotovora* (Jones) Bergey ssp. *carotovora* (Jones) Bergey. The incidence of soft rot was low at a K:N ratio of 4:1, increasing at 2:1 and 1:1. Verhoeff noted similar trends in tomato stems infected by *B. cinerea.* Paradoxically Verhoeff noted that high soil nitrogen can delay the development of latent lesions of *B. cinerea* in tomato, possibly because stem senescence is delayed. Over-luxuriant foliage is conducive to greater damage by sap-sucking insects such as aphids.

Pruning and Training

Pruning and training tall staked and wire-supported crops like peppers, tomatoes and cucumbers not only modify the microclimate by altering spacing but pruning alters the fruit:foliage ratio and hence source-sink relationships in photosynthates and the disease-susceptibility of various tissues. Removal of leaves bearing prepupal and pupal stages of pests can reduce their populations, but premature removal of leaves bearing parasitized stages can result in loss of biocontrol.

Fruit Load

Closely related to the management of pruning is the distribution of photosynthates in heavily cropping plants in relation to the susceptibility of tissues to fungal and bacterial pathogens. As Jarvis pointed out, modern technology has increased yields of greenhouse vegetables several-fold in the last two decades, with accompanying source-sink stresses on cultivars that have not changed very much. Thus, diseases such as Fusarium crown and root rot of tomatoes and Penicillium stem and fruit rot of cucumbers have become serious in that same period.

Both have been shown to be stressrelated and there has been a resurgence in the incidence of corky root rot of tomatoes that might be related to a diminished flow of photosynthates to roots. Grainger referred the "plunderable" carbohydrates available to certain pathogens – the so-called high-sugar pathogens– which include *B. cinerea,* whereas other pathogens, notably *Fusarium* spp., are classed as low-sugar pathogens principally attacking tissues starved of photosynthates. It is therefore incumbent on the grower to manage the nutrition, light and pruning of fruit and foliage so that a balanced partition of assimilates is attained without unduly compromising yield.

BIOLOGICAL CONTROL AND CROP DISEASE MANAGEMENT

Biological control is increasingly recognized as the foundation of a many successful IPM programs. Biological control has a number of benefits. It is environmentally benign and once established can give long-term control. After initial costs it is relatively inexpensive. It also ensures control over a large area compared to other control measures such as chemical spraying which only gives control in the area treated. It can contribute to healthy food, cleaner environment and increased profits for farmers. In Asia it has been demonstrated that outbreaks of the brown planthopper (BPH) in rice were caused by the destruction of effective natural enemies through the indiscriminate use of broad-spectrum insecticides. Failure to recognize the importance of natural biological control led to continuous outbreaks of BPH in several countries in Asia until the FAO Inter-Country Programme encouraged countries to appreciate this ecological phenomenon.

This Programme has encouraged extension staff to understand the role of natural enemies in their farming systems. This has been done by implementing Farmer Field Schools (FFSs) where biological control is the main focus of an integrated approach to pest control. Unfortunately, efforts at biological control of plant pathogens have lagged behind developments in the field of entomology and there are few examples of successful field implementation of biological disease control. However, significant development work would seem to imply that biological control has great potential for the sustainable management of plant pathogens. Biological control of plant pathogens usually refers to the use of one organism as an antagonist to another.

This can take the form of suppressive soils that contain antagonistic microorganisms, using antagonistic and trap plants. These biological control agents may be employed to keep the pathogen population below an acceptable threshold level, or may be used as a living barrier that prevents pathogen infection. Traditional farmers have been using biological control for millennia through the development of suppressive soils and this was highly developed in China and a number of Pacific Island countries. Examples of biological control of plant pathogens include the following: Bacteria such as *Bacillus cereus, B. subtilis* and fluorescent pseudomonads have also been successfully used to control fungal root pathogens. In Indonesia, fluorescent pseudomonads have been shown in the laboratory to reduce the severity of

Ralstonia solanacearum, the cause of blood disease in banana. Endophytic fungi of durian may have some potential as an antagonist of *Phytophthora palmivora*. In Taiwan, a *Bacillus subtilis* isolate was recently commercialized by Kuan-Hwa Chemical Co. for the control of powdery mildew of pea. *B. subtilis* was also found outstanding because of the superior antagonistic activity against *Rhizoctonia solani* AG4, *Sclerotium rolfsii*, and *Xanthomonas* sp. pv. *mangiferae-indicae* (Xcm).

Its mass production technique has been developed in Taiwan. Commercial preparations of *Trichoderma* and *Streptomyces* spp. are also commercially available. Inoculation of a healthy plant with a mild, or less severe strain, of a particular virus, gives protection to the plant against other more severe strains of the same virus. Cross protection technology has been used in some countries of the Asia-Pacific region. This includes work on cucurbits infected with zucchini yellow mosaic virus (ZYMV). The process involves mechanically inoculating the leaves of cucurbits with the mild strain of the virus before they are transplanted into the field. In one trial in the Pacific, cross protection resulted in a reduction of almost 50 per cent in the levels of disease compared to unprotected plants. Cross protection has been used in Taiwan and some other countries of Asia with mild strains of citrus triesreza closterovirus (CTV). One obvious problem with this technique is the need for labor to carry out the mild strain inoculations before plants are transplanted into the field.

Host Plant Resistance and Crop Disease Management

Traditional farmers in the Asia-Pacific region have been selecting, consciously and unconsciously, disease-resistant plants for centuries. These farmers often have knowledge of a large number of traditional varieties that they know will give good agronomic performance but also have a degree of resistance to the many injurious diseases that attack their fields. More recently, scientists have been breeding plants for improved resistance to plant disease. Since the 1960s, host plant resistance has been increasingly successful in the management of important diseases of staple food crops in developing countries with significant and sustained benefits to both smallholder farmers and their environment. The gains from resistant varieties have been most significant in Asia.

HYVs of wheat and rice of the Green Revolution with resistance to fungal and viral pathogens have contributed significantly to increased productivity, cheaper food and reduced poverty. Similar gains have been experienced in India with disease-resistant legumes. Banana BLS-resistant varieties released by the Honduran Agricultural Research Foundation (FHIA) breeding program have had some positive impact in Pacific countries. These FHIA hybrids are currently undergoing screening for resistance to other pathogens including nematodes and are being evaluated in over 50 locations. Agricultural Research for Developing Countries (CIRAD) have plans to screen bananas from their breeding program for BLS resistance in New Caledonia. In the Pacific, the TaroGen project has supported taro breeding programs in PNG and Samoa. These programs, based on durable resistance to TLB, have resulted in the release of nine new resistant varieties. There are plans to distribute these new varieties to other countries in the Pacific with TLB and also to countries still free from the disease. Resistance conferred on a plant by one or a few genes is often referred to as monogenic/oligogenic or race-specific resistance. This type of resistance usually results in the hypersensitive response and gives complete control against the pathogen.

This is easy to work with in breeding programs because it is easily incorporated into plants and gives a high level of resistance. Unfortunately it is this type of resistance that is easily overcome by new races of the pathogen. For this reason, resistance based on one or a few genes has generally not proved to be long-lasting. The breakdown of resistance in potato to *Phytophthora infestans* clearly illustrates this point. When many genes control resistance it is referred to as polygenic or non-specific resistance. Unlike specific resistance this type of resistance does not give complete control but the host plant is able to reduce or restrict the colonization or spread by the pathogen or limit the rate at which it can produce spores. Because none of the genes involved confer resistance if acting in isolation they are referred to as minor genes.

Non-specific resistance has the advantage of being long-lasting or durable, and it is being increasingly utilized in breeding programs such as the TaroGen taro breeding projects mentioned above. Robinson describes a number of innovative participatory approaches to plant breeding based on horizontal or durable resistance aimed at 'demystifying' plant breeding and taking control away from professional breeders and putting it in the hands of university students and farmers. He describes the formation of university breeding clubs and decentralized farmer participation schemes. The first university breeding club in the world was started in 1995 in Mexico. A university taro-breeding club was initiated at the University of the South Pacific (USP), Samoa in 1999.

This was the first club to be inaugurated outside of Latin America. The club represents an innovative approach to teaching and learning at USP. It is a cheap and easy approach to breeding. It ensures that there are many hands to do breeding work and has resulted in increased taro breeding activity. Robinson has proposed university breeding clubs as a "hands-on" approach for students to learn about breeding for horizontal resistance and a way of "scaling-up" farmer participation in plant breeding.

Robinson envisaged student members of breeding clubs returning to their family farms with potential new cultivars for evaluation. After a few decades, there could be hundreds, or even thousands, of former club members testing new lines as they emerge from clubs. Additional breeding clubs would increase the output even more, providing the widest extent and the highest possible quality of farmer participation in plant breeding.

BIOLOGICAL CONTROL

For the context of this paper, biological control is defined as the use, or encouragement, of natural enemies of a pest to reduce pest abundance. Four types of biological control are recognized, namely; classical biological control, inundative method of biological control, augmentation method and a management method. Classical biological control in Asia is exemplified by the first case study. This case study showed that it is not enough to introduce an efficient parasitoid. It is equally important that farmers are educated on the subject so that they can be equal partners in developing this nonpesticide method.

As shown by Ooi, farmers had to wait up to 12 years to realize the impact of the parasitoid. Lessons learned from this case study were transferred to the highlands of Dalat in Vietnam and cabbage farmers studied the biology of the parasitoid. Farmers also studied the use of a microbial insecticide (*Bacillus thuringiensis* [Berliner]) to replace chemical pesticides. When cabbage farmers in Dalat learned about the biology of the parasitoid and the action of *B. thuringiensis*, they stopped using chemical pesticides. The parasitoid was so effective that no further intervention was necessary. Inundative biological control involved rearing large numbers of biological control agents and releasing them into the field. Lim and Chong reported the periodic release of *Trichogrammatoidea bactrae fumata* (Nagaraja) (Hymenoptera: Trichogrammatidae) to control the cocoa pod borer. While it appeared to be able to reduce cocoa pod damage, many farmers find the process both tedious and expensive.

Moreover, rearing large numbers of parasitoids for timely releases can be very challenging. In some cases, the inundative releases appeared to have a placebo effect, allowing other natural enemies to control the pests when farmers stopped spraying chemical insecticides following periodic releases of the agent. In the case of augmentative biological control, beneficial insects are introduced at the time when pest populations are about to build up. An example of augmentative biological control is the encouragement of the black ant, *Dolichoderus thoracicus* (Smith) (Hymenoptera: Formicidae) to manage cocoa mirid pest. This involves establishing nests of the ant in cocoa plants. Liew also found that the black ants could control the cocoa pod borer.

Case study demonstrates that a larger biological control exists as a naturally occurring phenomenon. However, we often do not appreciate this and rice farmers were not aware of natural enemies in the field until they attended Farmer Field Schools (FFSs). Farmers studied and learned that friendly arthropods can keep a serious pest like the brown planthopper (BPH) in check. In the process, they soon discover that other pests like the rice stemborers are also kept at low populations by natural enemies. The most frequently asked question at rice FFSs is why farmers are not aware of such a ubiquitous phenomenon as natural biological control. It suggests that farmers were denied this information. It is ironical that farmers who live next to the rice fields are not aware of friendly insects that protect them from pest species. Hence, there is a need to educate farmers so that they can make more informed decisions about growing a healthy crop.

NATURAL BIOLOGICAL CONTROL OF THE RICE BROWN PLANTHOPPER

Before 1977, the rice BPH, *Nilaparvata lugens* (Stål) (Hemiptera: Delphacidae) was not even considered an important pest of rice in Malaysia. Like other countries in Southeast Asia, severe outbreaks demanded answers as to their causes.

Many suggestions were advanced but groundbreaking studies by Kenmore showed that disruption of natural biological control was the main cause. Key natural enemies were generalist predators such as *Lycosa pseudoannulata* (Boesenberg and Strand) (Araneae: Lycosidae) and *Cyrtorhinus lividipennis* (Reuter) (Hemiptera: Miridae).

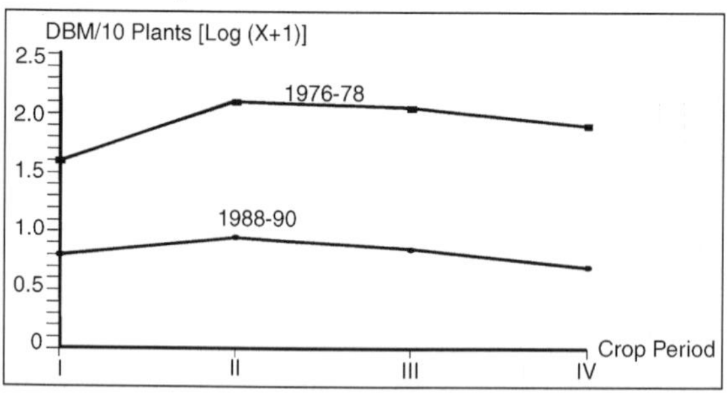

Fig. Seasonal Population Changes of DBM for 1976-78 (Pre-introduction of Diadegma Semiclausum) Compared with that for 1988-90

Repeated outbreaks of *N. lugens* in Indonesia was attributed to extensive use of insecticides, which were heavily subsidized by the government. Outbreaks were reduced following draconian measures taken by the government to ban 56 insecticides from the rice field and gradual removal of insecticide subsidy.

Insecticides killed off both pests and natural enemies and often pests recovered first leading to higher population levels than before, a phenomenon known as resurgence. Hence, in an area in Sekinchan, Malaysia where farmers reportedly treat their fields on a prophylactic basis, a field with three applications of insecticides had fewer predators as compared to fields with hardly any insecticide treatment. Higher ratios of BPH to predators (mainly *L. pseudoannulata* and *C. lividipennis*) in sprayed fields suggested that use of insecticides led to higher populations of BPH.

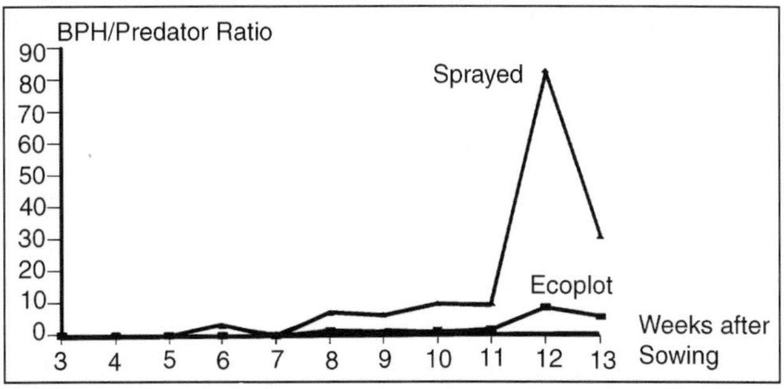

Fig. Number of BPH Per Predator (Mainly L.Pseudoannulata and C.Lividipennis) in an Ecological Plot (Ecoplot) and a Sprayed Plot

The importance of predators in maintaining low populations of BPH was also shown using exclusion cages. It was shown that in the absence of predators, BPH populations reached a very high level. In cages where the bottom was opened to allow predators to enter, both *L. pseudoannulata* and *C. lividipennis* helped maintained BPH populations at very low levels. This simple experiment can be carried out by farmers in FFS and is a very powerful tool to understand the impact of biological control.

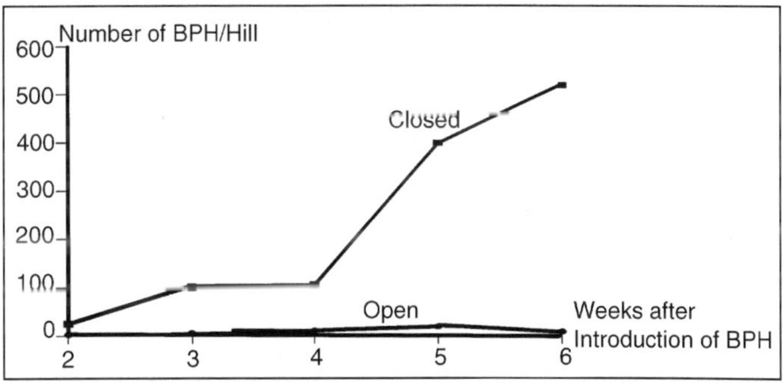

Fig. Results of an Exclusion Cage Experiment Set Up in Sungai Burong, Malaysia

PHYSICAL AND CULTURAL CONTROL OF INSECT PESTS

Prior to the availability of chemical insecticides, farmers developed various means to prevent insect pest infestation. Some of these have been refined through research both by farmers and researchers. Case study No. 3 is one such example of how farmers rediscovered an old method of avoiding pest attack. By studying the biology of the insect pest, farmers were able to confirm the findings made some 60 years ago by entomologists. More importantly, the information is now used by farmers to manage an important pest of rice. In Bangladesh, farmers who participated in vegetable FFS organized by an Integrated Pest Management (IPM) program funded by DANIDA (Danish International Development Agency) learned that the brinjal fruit and shoot borer, *Leucinodes orbonalis* (Guenée) (Lepidoptera: Pyralidae) pupated in leaf debris on the ground. Using this information, farmers would maintain clean fields and destroy all infested brinjal fruits. This dramatically reduced the incidence of the borer without using insecticides. In the past, farmers sprayed as many as 50 times per crop season! Cocoa is an introduced crop in Southeast Asia. The key pest is the cocoa pod borer, *Conopomorpha cramerella* (Snellen) (Lepidoptera: Gracillariidae). It is one crop that physical and cultural control received much attention.

Shah described several methods that were used to control the cocoa pod borer. These included bagging of all husks following extraction of cocoa beans and it was found that a minimum of nine days period was required to effectively kill all the larvae, pupae and adults. Besides bagging the husk, it was suggested that harvesting pods at the first hint of yellow may leave a large number of larvae inside the pod and bagging the husks will kill them off. To reduce pest population, it was recommended that a high standard of harvesting be maintained and that harvested pods be split the same day to extract the beans. Ecological harvesting refers to adoption of four aspects essential to effectiveness, i.e., regular harvesting, complete harvesting, a low minimum ripeness standard for plucking, and continuation throughout the year. A more draconian measure to manage the cocoa pod borer was reported by Roepke which involved complete removal of all cocoa pods and related fruits attacked by the cocoa pod borer, a method named "rampasan". If carried out efficiently, it would encourage a pest-free period the following season. For smallholders, the use of a transparent plastic bag to sleeve young pods has been practised in the Philippines and Indonesia.

IMPORTED PARASITOID MAKE DIAMONDBACK MOTH A NONPEST

A review of diamondback moth (DBM) management in the early part of 1970s would reveal that the literature was dominated by chemical evaluations for poisons to control this cabbage pest. The preoccupation with chemical control resulted in many reports of insecticide resistance. The response by the research community was to find more toxic chemicals with similar results – development of insecticide resistance. At that point, it was obvious that the only option for DBM management was chemical control and the frustrations of cabbage farmers drove them to use cocktails which increased the cost of production.

This suggests that farmers want to be involved in DBM management, which hitherto was left to researchers and scientists with very little change from the 1950s. An ecological approach was initiated in the 1970s in the main cabbage growing area of Malaysia. This served to widen the options available to farmers when some parasitoids were imported. However, these remain options until a chain of events resulted in a severe reduction in use of chemical insecticides in Malaysia. When farmers reduced the use of cocktails and focused on *B. thuringiensis*, the impact of *Diadegma semiclausum* (Hellen) (Hymenoptera: Ichneumonidae) was realized. However, farmers had to wait 12 years to appreciate this! The lesson learned is that farmers should be involved in biological control from the beginning. Educating farmers about ecology should take precedence in pest management program to ensure a sustainable program. Farmers would not stop using chemical insecticides unless they understand biological control. At the heart of this is the lack of appreciation of ecology. Farmers, hitherto, had been exposed only to a unilateral approach to pest control and it would take a great effort to unlearn this and understand the need for a more holistic approach.

ECOLOGICAL UNDERSTANDING OF THE WHITE STEM BORER

Despite annual outbreaks of the white stem borer, *Scirpophaga innotata* (Walker) (Lepidoptera: Pyralidae) in the district of Indramayu, province of West Java, rice farmers were not aware of nonpesticide methods for pest control. Equally intriguing was the lack of understanding by the authorities of studies to control this pest carried out some 60 years ago. It was not surprising that farmers did not understand the biology of the white stem borer, despite long association with the insect. Farmers observed that the white stem borer would occur in the first season following a three months drought. The farmers reported often seeing swarms of them in this first season.

The only approach adopted by farmers and recommended by agricultural authorities in 1995 was to spray with chemical insecticides. Despite numerous sprays, the rice crops were often devastated by the white stem borer. The IPM facilitator assigned to the study used this observation as an entry point to introduce studies that help farmers find out what happened. The IPM facilitator suggested that farmers carry out studies to determine the effects of insecticides. So, in July 1995, some white stem borer moths were collected from the fields and farmers sprayed them with insecticides. To their surprise, the moths spawned eggs before they died. Farmers then concluded that spraying insecticides did not prevent egg laying. The farmers' group in the village of Kalensari were graduates of IPM FFSs.

This group of 25 farmers decided to continue to work together to discover why they had the white stem borer problem. Regular meetings were held. Farmers went into the field to make observations and carry out field experiments. The IPM facilitator helped the farmers in experiment design and implementation. In one study, farmers evaluated the official recommendation of using an insecticide (carbofuran) coated on sand particles and applied in the seedbed. Results were variable, depending on the time of oviposition of the eggs. Hence, farmers realized that application of carbofuran did not guarantee an insect-free nursery. Armed with this information, farmers went further, exploring why there were still outbreaks despite spraying insecticides on egg masses. They collected freshly-laid egg masses and sprayed these using the recommended dosage. Farmers noticed that after about a week, healthy larvae emerged. The results convinced farmers that spraying insecticides really did not stop an outbreak. They had to look for another approach and subscribed to the principle that the more they understood the biology and ecology of the stemborer, the greater the potential for management. The next study concerned learning about the origin of early season flights. Farmers suspected that the moths came from the fields and decided to survey rice stubbles left in the field from the previous season.

To their amazement, they found live larvae within some of the stubbles. Discussions with resource persons helped them understand the concept of insect aestivation (a form of hibernation during adverse weather conditions), observed by Goot. This discovery led to different experiments to kill off the aestivating larvae. This showed that burning and flooding did not kill many larvae, and that the best approach was to remove egg masses. They organized a local campaign to do this. This campaign proved successful, as the incidence of stem borer in the locality was about 5 per cent compared to neighboring areas which had about 25 per cent bored tillers with prophylactic spraying of chemical insecticides. Continued discussion among farmers led to the design of experiments for avoiding the peak flights of the stem borer moths. Farmers carried out studies to understand when the stem borer moths took flight. They were able to duplicate Dammerman's study with light traps, using them to monitor the moth flights. Farmers also learned the role of water in breaking aestivation. Combining this information, farmers set up a strategy to 'forecast' peak flights and organized other farmers in their villages to only plant nursery beds after peak flights.

A PARTICIPATORY IPM APPROACH TO ENCOURAGE NONPESTICIDE METHODS

The problems arising from use of chemical insecticides were realized as early as 1950s. By the early 1950s, resurgence of pest outbreaks and resistance to chemical insecticides were reported from all over the world. Initial attempts led to the first development of IPM as a combination of chemical and biological control. This did not go far enough to address the conflict of using two opposing approaches.

Chemical insecticides by their very mode of action remove or inhibit the function of biological control and this poses a problem of integrating chemical and biological control. Recognizing this, steps were then taken in the 1960s to formulate a more ecological approach. In the 1970s, the approach recognizes the need to identify social and economic requirements in addition to ecological consideration. Despite all attempts at defining IPM, it remained very much at academic levels, particularly in the wet tropics. Part of this problem lies in the heavy reliance on the use of economic thresholds as a key element to decision-making. Economic threshold levels were insensitive to both social and economic situations of farmers and it was realized by 1980s that the use of economic threshold levels continued to maintain unnecessary use of insecticides due in large part to lack of understanding of underlying ecological principles. Realizing this dilemma, the FAO implemented IPM program for Asia in the late 1980s initiated a nonformal education approach to help farmers understand these underlying ecological principles. The process came to be known as the IPM Farmer Field School. The IPM FFS is the primary learning approach piloted in Indonesia.

The IPM FFS is a season-long learning experience. In the FFS, farmers learn about agro-ecosystem management. The FFS makes use of the crop field as a field laboratory. In the laboratory, FFS participants learn about the ecology of the crop by means of regular observation and hypothesis testing. Discussions are based upon participant analysis of field conditions. The FFS uses a participatory learning process. The process emphasizes taking decisions and actions based on an open discussion of ideas which is free from the domination of any individual. These decisions are tested in the field laboratory. The FFS process, besides its emphasis on field ecology, provides participants with an opportunity to examine human social dynamics. As a result, FFS participants not only learn about the cause and effect relationships that exist in the field; they also acquire a greater understanding of human relationships. The analytical processes employed in the FFS enhance farmers' capacities to examine the conditions, in which they live and work. Participants, having completed their FFS, are able to take decisions and take actions that would improve those conditions. The increased understanding of participants regarding human social dynamics enables them to develop collaborative efforts to ensure that planned actions are implemented. Even after the FFS, farmers continued to experiment and in some cases worked in groups. For example, farmers in the village of Kalensari continued to study a nonpesticide method to control the white stem borer. Examples of follow-up studies by farmers are reflected. This confirms that farmers are able to carry out experiments and develop innovative nonpesticide methods of controlling pests. As noted in the first case study, encouraging IPM farmers to continue research is critical to the realization of the potential of a biological control agent. Results from a better understanding of nonpesticide methods have helped farmers to convince policymakers to take the right steps in preventing outbreaks. It is time that policymakers should make this approach *a priori*. An enabling policy to encourage farmer education in FFS and carry out field ecological studies will invariably promote the development of more nonpesticide methods to manage insect pests in Asia.

ISSUES CONSTRAINING THE USE OF NONPESTICIDE METHODS OF CROP DISEASE MANAGEMENT

The continued promotion of nonpesticide methods for crop disease control is essential for the development of a more sustainable agriculture. However, there are important obstacles that at present minimize their impact. One of the most important factors is the power of the pesticide industry and its lobbyists often reinforced in countries that provide subsidies for chemicals. Even if subsidies are not provided directly they are through the costs required to treat those whose health is compromised by their use or adverse environmental impact. Governments have a significant role to play here by stronger regulation of the pesticide industry.

This would include restriction on the use of very damaging pesticides and a price system that is a true reflection of their real impact. At the same time governments need to actively support the use of nonpesticide methods through increased allocation of resources for research and staff and a policy environment that encourages the use of nonpesticide methods of control. More specific issues that need to be addressed at the individual method level are outlined below. Within the Asia-Pacific region there is a wide discrepancy in the capacity of national quarantine services. Most, if not all, quarantine services in the region are constrained by one or a few of the issues that follow. Many quarantine services in the region are under-resourced in terms of trained human capacity and modern technologies required to effectively police the movement of plant material.

This is more the case for Pacific countries but it applies also to many of the Asian countries that have much greater levels of trade and commerce to deal with. Many Asian countries share a common border and have multiple entry points that require more demands from quarantine services. This limited capacity makes it difficult for countries to adequately respond to a disease incursion and achieve eradication. Good quarantine requires effective regulations and laws. Although most countries in the region have adequate legislation there are still a few countries that do not. Pathogens pose a serious challenge to plant quarantine services compared to other pests such as insects because they may be introduced as spores or eggs on unsuspected carriers or occur as latent infections. Plants or plant parts with virus infections may not show symptoms.

Very often quarantine services do not have adequate personnel trained in disease recognition or the necessary serological/molecular diagnostic tools. In the case of the Pacific, few of the major pathogen problems have been adequately characterized such as viruses causing disease in taro, kava and yams. Sensitive diagnostic methods have not been developed and this seriously constrains the international movement of crop germ plasm. This has been a major constraint to crop improvement programs in the Pacific. An important dilemma facing all countries of the region is the urgent need to find a viable alternative to methyl bromide which is to phase out by the Montreal Protocol in 2005. Certification schemes or programs to produce pathogen-free planting material are costly and require a certain level of infrastructure and human capacity. This is required to enforce regulations and ensure a degree of quality control. Schemes are also reliant on access to sensitive diagnostic methods. Not all countries in the region have this capacity. In some countries, e.g., Thailand, there was a perception among most farmers that disease-free plants were resistant to pathogens when grown in orchards. This indicates the need for education and awareness to accompany programs and the diseases being targeted. What do farmers need to know to ensure that pathogen-free planting material remains so for longer? Other constraints identified in such programs include difficulty in getting farmers to remove healthy-looking trees and plants that are infected (again, education is very important).

There may be a requirement to use pesticides to control vector buildup to ensure the program is effective. Molecular diagnostics are expensive and require expertise that may not be available in some countries. There is a need for simple and cheap diagnostics. The effectiveness of such certification strategies is dependent on the ability to accurately diagnose and detect viral pathogens. In certain countries, such as Thailand and Malaysia, there are problems related to accelerating the production of pathogen-free citrus seedlings to meet demand. Although cultural methods offer appropriate approaches to obtaining sustainable disease management and a sound alternative to pesticides it is important to remember that many of the methods are localityspecific. Some methods are also labor-intensive and generally have not become widely adopted by farmers due to economic constraints and the long-term nature of the methods.

The benefits and advantages of biological control are well documented. However, there are relatively few successful examples of the widespread application in disease management. There have not been the same successes as have been achieved with insect management. Despite extensive research on the use of biological control of plant pathogens there are few practical applications, even fewer for the Asia-Pacific region. Many of these antagonists are already present naturally in the soil and it may be best to encourage their activity through some of the other methods mentioned and which are common to traditional farming systems. Host plant resistance has been an important tool for achieving sustainable crop disease management. Despite this it is a relatively long-term activity.

In addition, many countries in the Asia-Pacific region do not have the capacity to sustain breeding programs. At present there are only two breeding programs in the Pacific on taro and rice. The issues surrounding the use of biotechnology and its application to agricultural improvement are many and complex. There are many technical issues, such as the development of gene delivery and transformation systems, that are constraining developments for specific crops but these will be overcome with ongoing research. Although biotechnology and its role in disease management appears to be progressing in Asia, it is worth discussing its possible role in the Pacific region. It has already been highlighted that Green Revolution technology had a minimal role in the Pacific region and unless things change considerably over the next few years it is possible that the technology gap between Asia and the Pacific will widen further. At present there is little capacity or resources to undertake biotechnology research in the Pacific.

There is little that exists in the form of policy or a regulatory framework to facilitate biotechnology research or applications. Costs and resource requirements for biotechnology are high. Can Pacific countries afford the costs required to undertake such research? If they cannot, can they rely on biotechnology organizations, private or public, to address their relevant crop disease needs? If they can, will farmers be able to afford the end-product? New transgenic plants are the product of millions of dollars in investment. The price of the end-product is likely to be beyond what a typical farmer in the Pacific can afford to pay. Staple crops important in the Pacific have been neglected by past developments in crop protection technologies, will they continue to be? Who are these neglected crops going to interest? Some of these issues apply to countries in Asia with similar human capacity and resource constraints.

The other issues that concern the role of biotechnology are more general and have been summarized by Conway and Tripp as:

- *Environmental issues*: Gene transfer to wild relatives (genetic pollution), impact on natural fauna and flora, pathogen resistances.
- *Health issues:* Possibility of transferring antibiotic resistance to soil fauna and eventually to livestock and humans, allergens ('Frankenfoods'), toxins, long-term health effects.
- *Ethical issues*: Naturalness of genetic engineering, bypassing of public interest, domination by a few major private companies.
- *Benefits to the poor*: Are the poor going to benefit, will farmers and consumers in the Asia-Pacific benefit?
- *Sustainable agriculture issues:* What will be the impact on the pursuit of sustainable agriculture, will biotech lead to increased domination by large-scale agriculture? There is concern regarding the 'marriage' between the pesticide and biotechnology industries.

STRATEGIES FOR FACILITATING THE SUCCESS OF NONPESTICIDE METHODS

Countries of the region need to appreciate the real cost associated with the past use and reliance on pesticides for pest control. These costs have been highlighted in depth in this paper. There is an urgent need to actively support and promote nonpesticide methods if we are to change from the prevailing approach to pest control to a more ecological approach.

It is recommended that:

- Governments put in place policies to promote all aspects of nonpesticide and a more ecological approach to crop disease control;

- Farmers be educated and trained to enhance their skills and knowledge of nonpesticide methods. Over the last decade, there has been a significant shift in the emphasis of IPM programs throughout Asia, towards a more participatory, decentralized, community-based approach, termed community IPM, in which the farmers become the initiators, implementers and promoters of IPM and not just the recipients. In community IPM, farmers organize, manage and implement their own IPM activities, analyze problems, design field studies and carry out experiments and undertake efficient farming practices. This involves a range of approaches such as FFSs and farmer-to-farmer training, considered to be a costeffective, decentralized, community-based, farmer-first approach that will promote sustainability and expansion of IPM and reduce reliance on pesticides. Although community IPM and FFS initially grew out of problems related to insect pest management there is a growing awareness among plant pathologists of their relevance to sustainable crop disease management in the region. It is recommended that the FFS approach continue to be explored to enhance the use of nonpesticide methods;
- Linkages are developed between all stakeholders (farmers, extension workers, researchers, NGOs and other community-based groups including consumers) to enhance awareness and knowledge of nonpesticide methods through more effective networking;
- Countries strengthen regulation of the pesticide industry to reflect its real cost;
- There is further cooperation and coordination among national/ regional/international organizations and donors to highlight the problems with pesticide use and to promote a more ecological approach to crop disease control.

NONPESTICIDE METHODS FOR MANAGING CROP WEEDS

India produces 201.6 million mt of food in 124 million ha with a productivity of 1,625 kg/ha. The oilseed production is about 22.3 million mt grown in 25.26 million ha with a productivity of 873 kg/ha. Pests and diseases cause huge damage in the production of various crops. It is estimated that 53.1 million mt of food grains and 56.8 million mt of sugarcane worth ₹.292.4 billion were lost to pests and diseases attack.

Normally the farmers depend on chemical pesticide for the control of pests and diseases. The use of chemical pesticides has caused a lot of environmental hazards apart from resistance developed by the pests to the chemical.

In view of the above reasons, nonpesticide methods of controlling pests and diseases have gained considerable importance.

STATUS OF DIFFERENT NONPESTICIDE METHODS

Mechanical Methods

The moths of several harmful insects emerge during *Kharif* (June-September) season as well as in subsequent months. After the emergence, male and female moths mate and gravid female lay eggs from 500 to 2,000 in soil or on crop plants. These moths normally rest under clods, in cracks or under dry vegetation matter in soil. Such moths can be physically collected and burnt. This helps in reducing population of pests. Collection and killing or trapping of moths of red-headed hairy caterpillar, *Amasacta albistriga* Walker (Lepidoptera: Arctiidae), cotton bollworms, *Earias insulana* Boisd. (Lepidoptera:Noctuidae) and *Earias vitella* Fabr. (Lepidoptera:Noctuidae), pigeon pea and chickpea pod borer, *Helicoverpa armigera* Hübner, (Lepidoptera:Noctuidae) have been reported from the endemic areas of Karnataka, Tamil Nadu, Kerala and Andhra Pradesh in South India. For instance, in Pavagada taluk of Tumkur (13º18′ N, 77º12′ E) and in Karimnagar (18º26′ N, 79º10′ E) and Mahbubnagar (16º45′ N, 77º59′ E) districts of Andhra Pradesh in South India, villagers collect the moths and caterpillars in large numbers, deploy light-traps for attracting moths and mass-kill the insects. Apart from the adults, the larvae or grubs in soil can be collected large numbers and destroyed.

The root grub comes out of the soil during summer (February-April) or pre-monsoon (May-June) showers and feeds on the leaves or fresh twigs of neem (*Azadirachta indica)*, tamarind (*Tamarindus indica*) and *Ficus* species. Adults of root grubs are attracted to light and bonfire. Such adults can be collected in containers containing kerosene and water and destroyed. It is an important, proven integrated pest management (IPM) tool, practiced in many South-East Asian countries. Mass collection and destruction of root grubs has been reported from endemic areas of Karnataka, Gujarat, Maharashtra, Rajasthan, etc. For instance, in Karnataka, sugarcane (*Saccharum officinarum*) root grubs are collected in large numbers and destroyed in Bhadravathi taluk of Shimoga (13º57′ N, 75º32′ E), Gundlupet in Mysore district (13º15′ N, 77ºO′ E), of coffee (*Coffea arabica* and *Coffea robusta*) in parts of Chickamagalur (13º15′ N, 75º45′ E) district, of areca (*Areca catechu*) in Thirthahalli, Sagar and Soraba taluks of Shimoga, and of groundnut (*Arachis hypogaea*) in Chintamani taluk of Kolar district (13º12′ N, 78º15′ E).

The practice of collecting and destroying in huge numbers the eggs, grown-up larvae and adults of pests like grasshoppers, locusts, hairy caterpillars on bunds, sticks, weeds, stretches of wastelands, is common. Such insects can be collected and destroyed or put in a tin pasted with grease on the top portion. Such larvae are killed in tin. *H. armigera* is a pest which is difficult to suppress by methods other than mechanical methods. The larvae can be collected by stirring the plants and by sweeping the larvae fallen underneath the plants. Such larvae are concentrated in pits and can be destroyed/burnt. Sugarcane setts treated in hot water for 10 minutes controls whip smut (*Ustilago seitaminea*) in Gujarat. It is a recommended practice in all sugarcane areas in India.

Soil solarization in the vegetable nursery with the help of polythene mulch (25 microns) for 40 days after flood irrigation kills *Pythium* spp. and *Fusarium* spp. up to 30 cm soil depth. Raising vegetable nursery later in such solarized beds will have lower incidence of pest emergence, damping off, higher seed germination and better seedling vigor in Himachal Pradesh, Northwest India.

Clipping of terminal shoots that serve as oviposition sites for 110 days in cotton (*Gossypium hirsutum* and *G. arboreum*) and 40-50 days in chickpea (*Cicer arietinum*) reduces the *H. armigera* infestation in egg stage, which is being practiced in cotton, as well as in chickpea. Wooden or metallic 'T' or 'Y' perches are placed randomly in the cropped area to attract predatory birds to devour larvae/adults of pest insects. In India especially insectivorous birds like drongo (*Dicrurus adsimilis*) and mynah (*Acrido theres*) have been reported to be effective agents of suppressing populations of *H. armigera, Achaea janata* Linn. (Lepidoptera:Noctuidae) and hairy caterpillars in pigeon pea, chickpea, cotton, castor (*Ricinus communis*) ecosystems. Birds feed on the larvae and reduce their population. For instance, about 80 per cent of *H. armigera*. Spreading boiled rice balls or balls of Bengal gram preparation in semi-looper (*A. janata*) affected castor fields help in attracting birds, which eat larvae of semi-looper and reduce the populations. Spreading fried rice (puri) in *H. armigera* affected fields of chickpea also attract birds which predate on larvae of *H. armigera* and other pests.

Physical Methods

Light traps in the form of gas lights or mercury lamps placed in large contiguous areas help in trapping moths of *A. albistriga*, wherever the pest is endemic. This will reduce the moth population and multiplication. This has been reported from endemic areas like Pavagada taluk in Tumkur, Kadur in Chickamagalur and Hosadurga in Chitradurga district (14º15′ N, 76º28′ E) Karnataka, South India.

Cultural Methods

Monocropping, ratooning and growing the same crop year after year favors the pest and disease buildup. Therefore, it should be avoided. The population of the fungus, *Fusarium* sp. can be reduced when soil is kept fallow, compared to growing successive pigeon pea in the same soil, pigeon pea-fallow-chilies and pigeon pea-fallow-tobacco cropping pattern appreciably reduces *Fusarium* spp. Growing maize, (*Zea mays*) chickpea crop sequence decreases root-knot nematode population in chickpea (All India Coordinated Research Project [AICRP]).

Fall ploughing with a furrow turning plough by the end of the cropping season is advocated to expose the diapausing insect stages in several agro-ecosystems. The tillage practices expose the larvae and pupae of pest insects like cutworms (*Agrotis* spp.), hairy caterpillars (*Amsacta* spp.), grasshoppers (*Hieroglyphus* spp.), root grubs (*Holotrichia* spp.) and locusts (*Locusta* spp.) either to inimical conditions or to predators like insectivorous birds.

Planting time has been exploited by many entomologists and pathologists to check insect-pests and disease buildup. For instance, sowing black gram (*Phaseolus mungo*) in the second fortnight of June reduces the foliar diseases in the hill region of Uttar Pradesh. The pigeon pea fusarium wilt is reduced when the crop is sown after July and also by sowing under rainfed conditions in Kanpur, Uttar Pradesh, North India. Sowing early and synchronizing preferably with short duration genotypes lowers bollworms (*Earias* spp.) incidence and damage in cotton in Southern India. Planting rice during second fortnight of July reduced rice neck-blast in Haryana, India. Sowing mustard in October reduces white rust incidence in Karnataka.

Inter- and intra-row spacing has direct and indirect influence on the population of bollworm (*Earias* spp.) because of variation in plant habit and in architecture. Closer spacing (75 × 30 cm) attracted higher numbers of bollworms than wider spacing (90 × 30 cm). Wider spacing alters the microclimatic conditions to make it unfavorable for pest multiplication and also make easy to reach pest with bioagents and chemicals. Closer spacing of 25 × 10 cm reduces mung bean yellow mosaic virus compared to 40 × 10 cm.

Trap cropping is yet another technique to successfully suppress pests or pathogens in cultivated ecosystems. Planting highly preferred crops in and around the cotton field serves as trap crops for bollworms. Variation in attractiveness of crops for egg laying or larval feeding is exploited as an effective tool to concentrate early pest populations on selected plants. The major benefit of this is that either the insect can be collected manually or control measures with bioagents or chemicals can be effected, to a limited area, thus avoiding chemical applications to the entire area. The overall application will be less than in conventional farming.

Okra (*Ebelmoschus esculentus*), marigold (*Tagetus* spp.), pigeon pea, chickpea, tomato (*Solanum esculentus*), tobacco (*Necotiana tuberosum*), maize (*Z mays*), sorghum (*Sorghum vulgare*), and beans (*Phaseolus* spp.) have been used as trap crops in India and proved effective to lower the incidence of *H. armigera* on cotton. Several studies have been made in India to suggest that okra and marigold have the potential to work as effective trap crops for *H. armigera* on cotton. Trap crops may also attract natural enemy by the presence of semio-chemicals, thereby enhancing effectiveness of natural control. Suppression of diamond back moth (DBM) (*Plutella xylostella* Linn.) (Lepidoptera:Yponomeutidae) by trap cropping is popular among cabbage (*Brassica oleracea*) growers in and around Bangalore, Karnataka. For suppressing DBM, paired rows of mustard (*Brassica juncea*) for every 25 rows of cabbage planting and second row sown 25 days after planting exhibited 90 per cent attack to mustard than cabbage. Similarly planting marigold one row for every 16 rows of tomato attracts *H. armigera* moths to oviposit on marigold (*Tagetes* spp.).

Intercropping with cowpea, (*Vigna unguiculata*), maize, onion, finger millet (*Eleusine coracana*), mung bean, soybean (*Glycine max*) groundnut in south and central zones of India to colonize biological control agents have been advocated. Growing cotton with groundnut/green gram at 1:3 or with onion 1:4 or five proportions have been agronomically accepted practice for remunerative yield in Karnataka. Intercropping barley (*Hordeum vulgare*) with chickpea reduced blight disease (*Ascochyta rabiei*) in Rajasthan, India, marigold or cowpea planted around cotton crops as border crop reduces incidence of bollworms (*Earias* spp. and *H. armigera*) in cotton.

Manuring and fertilization are practices adopted to promote rapid growth in plants and shorten the period of susceptible stage of attack to the herbivores. Nutrients provide greater tolerance and compensates for insect damage. However, such practices may also change the physiology of plants in favor of insects to make the plant most desirable for survival and multiplication.

Selected examples are provided below for the role of nutrients in conferring resistance/susceptibility to insect attack or pathogen invasion:

- Increased nitrogen level lead to increased oviposition and increased infestation and damage by *H. armigera* in cotton.
- Use of 100 kg nitrogen/ha induces moderate incidence of false smut in paddy compared to higher doses of nitrogen in Gujarat.
- One hundred kg potash/ha decreases stem rot disease (*Macrophomina phasiolina*) and increased fiber yield in jute in West Bengal.
- Boron application at 2.5 ppm in pearl millet (*Pennisetum typhoides*) reduced the primary incidence of downy mildew to the extent of 76.4 per cent at 35 days after sowing in Rajasthan.

Managing or altering the moisture levels or extent of moisture in cultivated tracts is important for the management of pests and diseases as illustrated in instances given below. Early termination of irrigation avoids crop prolongation and this reduces bollworm succession. However, this is mainly for pink bollworm, *Pectinophora gossypiella* (Lepidoptera: Gelechiidae). Increased soil moisture due to mulching, farm yard manure and low plant population alone or in combination in cluster beans (*Cyamposis tetragonoloba*) helps to increase the population of resident bacteria, which reduces the dry rot caused by *Macrophomina phasiolina*.

Plant Resistance

Host plant resistance is deployed as unique approach for pest management with predictable problems. It helps in suppressing the pest population at low cost with least disturbance to the ecosystem. Unequivocal advantage of plant resistance is its compatibility with all other methods of pest control. Even low levels of field tolerance/resistance are desirable in cultivated plant varieties.

BIOLOGICAL CONTROL METHODS

Several species of natural enemies including parasitoids, predators and pathogens have been recorded in paddy, cotton, pigeon pea, chickpea, tomato and cabbage ecosystem. These natural enemies help in killing the eggs, larvae of different pest. These natural enemies play an important role in the agro-ecosystem.

Biorational Tools

Pheromones are semio-chemicals secreted by insects to complete normal life cycle activities. Entomologists have exploited these chemicals to interfere with the normal activities such as mating and reproduction and thus reduce their populations in the cultivated patches. Pheromones are also key component of cotton, pigeon pea, chickpea IPM and are mostly utilized for monitoring the pest populations so that timely interventions are made. Monitoring helps in early detection of bollworm populations and its buildup. Use of pheromones has helped in monitoring *H. armigera* population on tomato.

Two sex pheromone traps per acre were fixed and lures have to be changed in 15 days for monitoring *H. armigera* which helps in assessing attack of pests in advance and intensity of pest which helps to take control measures.

The important contribution of biotechnology is the capacity to express pesticidal proteins within transgenic plants. Bt cotton could provide effective control of target pests. Transgenic bollworm (*H. armigera*)-resistant cotton is seen as an improvement over traditional cotton and has achieved significant social and economic benefits. The use of insecticides in Bt (*Bacillus thuringiensis*) cotton decreased by 60-80 per cent and the density of predators increased by 24 per cent. The Government of India has given provisional approval for commercial cultivation of Bt cotton in March 2002.

SUCCESS STORIES

In Tumkur district of Karnataka during 2001-02, the Department of Agriculture organized mass collection of eggs, larvae and adults. An estimated 341 kg of moths were collected and burnt. Three thousand moths weighed 1 kg. Similarly eggs were collected in 2,000 ha. The eggs were collected and put in a tin coated with grease. Such eggs have been destroyed. During 2000-01, in the same area 65,728 mt of larvae were collected and burnt.

- Government sanctioned special package for the purpose.
- Action plan developed involving farmers, people's representatives,

the press, scientists, NGOs, Rotary Club members, merchants, media persons, etc.

- Awareness was created through street plays, video cassettes were shown, colorful wall posters were displayed, wall paintings were done, meetings were conducted in villages, banners were put at popular junctions, publicity was made through mike-systems, visits were made to areas where operations concerning hairy caterpillars were successfully carried out.
- Active groups were formed in each village.
- Distributed 2,300 gaslights under subsidy program for using in light traps.
- Set up 30 monitoring centers.
- Set up 200 mercury lamps for light traps.
- Clear-cut responsibilities were fixed and on works to be done.
- With rainfall, follow-up action were taken.
- Seeds of pulse crops were distributed to be grown as intercrops under subsidy schemes.

Root Grub *Holotrichea serrata* Linn. (Coleoptera: Scarabaeidae)

In Bellary, Dharwad district of Karnataka during 1986-87 and 1987-88 collected 20 mt of root grubs in each place after the receipt of summer showers in May 1986 and May 1987. The farmers were briefed through meetings about the pest, identified the endemic area with the help of farmers, conducted night campaigns during summer showers rainfall, light traps, bonfires with crop stubbles and bicycle waste tires were set up at night. The adult root grubs were attracted to light traps and fire. The beetles were collected and put in water mixed with kerosene.

Farmers Field School (FFS)

During 2000-01, with the help of Food and Agriculture Organization (FAO) technical assistance, 22 agricultural officers were trained as IPM facilitators. The training of facilitators was held at the Regional Research Station, Raichur, in Karnataka, South India. During 2000-01, 20 FFSs were conducted with the help of above IPM facilitators. The school was conducted in the farmer's field only for 20 weeks. Every week, the farmers gathered for the field school and learned to analyze the cotton agro-ecosystem, determine the population levels of parasitoids and herbivorous insects. They carried out experiments and understood the function of useful insects. Based on their understanding of the ecosystem, farmers were able to make informed decisions on growing a healthy crop. The following suggestions were evaluated by farmers to determine if they are useful:

- Okra, marigold, pigeon pea and cowpea were grown as trap crop to control sucking insects, *H. armigera* and *Erias* bollworm.

- Seed treatment with *Trichoderma* at 4 g/kg and with imidachloprid 7.5 g/kg.
- Stem pasting with imidachloprid @ 1:20 ratio to control sucking insects.
- *Trichogramma* released 100,000/acre to parasitize eggs of *H. armigera*.
- NPV spray @ 200 LE/acre.
- Set up 25 bird perches/acre for control of small larvae of *H. armigera*.
- Caterpillars were collected and destroyed.
- In 40 days old cotton crop, pheromone traps at 2/acre were set up to monitor the intensity of *H. armigera* and *Earias* bollworms.
- Twenty-five yellow sticky tins were set up to control whiteflies.

Farmer would usually apply 19 chemical sprays to control pests and diseases in their fields and spend ₹.9,000 whereas the cost of protection was only ₹.2,097 in IPM plots. The yield obtained in farmer's plot where IPM was not practiced was 600 kg/acre, earning ₹.9,000 at the cost of ₹.12,936 whereas in IPM plots, the farmers were able to get 1,100 kg of yield by investing ₹.6,043 and fetch the net returns of ₹.10,468 as profit.

In Tumkur district, Karnataka during 1995, one officer participated in a season-long training to become an IPM facilitator with the technical assistance of the FAO. Four FFSs were conducted for 10 weeks in which 20 officers and 120 farmers participated. Every week the farmers would assemble for half a day in the study field. They researched on parasitoids and other useful insects including pests using insect zoos. From the research conducted by farmers, they learned more about useful insects and this helped them to conserve the beneficials by not using insecticides. The farmers after this exercise did not use even a single chemical spray whereas in control plot the farmers had sprayed four times. The FFS concept is being continued in different areas every year.

Nonpesticide Method of Controlling Pests and Diseases in Chickpea

In 2001-02, 50 ha demonstrations were conducted at Chitradurga district, Karnataka, involving 16 farmers.

The farmers were shown the following practices:

- Seeds treated with 4 g/kg *Trichoderma*.
- Sex pheromone traps were put @ 2/acre.
- Nipping of tips as soon as eggs were noticed.
- Churumuri (roasted rice) spreading at 6 kg/acre to attract birds, which predates on *Helicoverpa* larvae.
- Erection of 10 bird perches/acre to attract birds that feed on caterpillars in the crop.

This has resulted in 750 kg/ha yield against 500 kg/ha in the control. In addition to regular cost of production, an additional cost of ₹.470 was spent on nonpesticide method for controlling pests and diseases. In farmer practice,

₹.1,200 was the additional cost for purchase of chemicals to control the pests and diseases. Hence, use of nonpesticide methods result in 250 kg of additional yield, which fetched about ₹.4,000 per net profit.

INTEGRATED PEST MANAGEMENT

Indiscriminate and injudicious use of chemical pesticides in agriculture have resulted in adoption of IPM as cardinal principle and main plank of plant protection in the overall Crop Production Program. IPM is an eco-friendly approach which focused on farmer education to make farmers more efficient in their production activities. The IPM concept is being promoted through 26 Central Integrated Pest Management Centers (CIPMC) located in 22 States and one Union Territory aiming at human resource development, these centers are imparting training to extension functionaries and farmers in IPM skill by conducting FFSs. In addition, these centers are engaged in mass-production and field releases of biological control agents. With a view to provide technical knowledge to the extension functionaries and farmers in the States, IPM approach has been tried out for 51 crops like rice, cotton, vegetables, pulses, plantation crops, spices, fruit crops and oilseeds. A participatory IPM approach based on farmer education is being implemented throughout the State by the extension service.

Towards human resource development, facilitation skills have been developed through continuous education of Agriculture Extension Officers (AEO) and farmers. IPM FFSs are conducted in the States to enhance the skills of farmers.

Aiming at organic farming production technology for biocontrol agents, biopesticides have been evolved which are being used by the private as well as public sector units for production of biocontrol agents to be used by farmers to save their crops from the ravages of pests and diseases. This concept prevents dependence on chemical/pesticides for plant protection. Since the inception of IPM program, 7,113 FFSs have so far been conducted thereby training 32,786 AEOs and 212,870 farmers in IPM skills. Eleven hundred and thirty-two IPM facilitators have been trained by conducting 35 season-long training of facilitators courses on cotton, rice, vegetables, pulses, oilseeds, etc. With the background of IPM knowledge farmers are practicing farming with minimum use of pesticides as agrochemicals.

Instead they are depending on biological control agents. As a result of implementation of IPM program, there is a significant reduction in the consumption of pesticides from 61,357 million mt (technical grade) during 1994-95 to 43,584 million mt (technical grade) during 2000-01. In the case of red-headed hairy caterpillar the government has given sanction to take up effective awareness campaign, providing gaslights under subsidy program, providing funds for catching larvae at ₹.35/kg and moth at ₹.50/kg. Farmers contact centers have been started at hobli level to deliver the different nonpesticide/IPM methods to farmers. Government has given permission to grow commercial Bt cotton in farmer fields in March 2002.

IMPACT OF VIRUS-FREE SWEET POTATOES

Sweet potatoes are currently grown by small farmers on 600,000 ha in Shandong province, China with an annual production of about 17 million mt. In the late 1980s, observations indicated high levels of virus infection and propagation of virus-free seed was identified as a possible means to improve yield. International Potato Center (CIP) and Chinese scientists began a project to develop and transfer methods for propagating virus-free planting material using tissue culture and Enzyme-linked Immunosorbent Assay (ELISA). Between 1994 and 1998, virus-free plants were extended to about 80 per cent of the hectares and contributed to a 30-per cent yield increase with no change in the use of inputs. The internal rate of return estimated at 202 per cent with a net present value of US$550 million. By 1998, yearly productivity increases were put at US$145 million, improving the income of the provinces seven million sweet potato growers by 3-4 per cent.

Compared to seed, vegetatively propagated planting material can potentially carry all pathogens that are present systemically in the mother plant in addition to pathogens that are carried externally. Pathogen-free plants may be obtained by meristem tip culture of the upper millimeter or so of the growing tip of the plant. Most viruses do not invade this portion of the plant for a number of possible reasons. Meristem tip culture is sometimes accompanied by heat treatment. This may involve putting plants in controlled temperature cabinets up to 40 ºC for a number of weeks. Under these conditions the rate at which new meristematic tissue is synthesized may outstrip the rate of viral replication. Shoot-tips removed from these plants should be virusfree. Virus and virus-like diseases are serious constraints to tropical fruit production in the region. Several large citrus development programs in the region have failed because of the quality of planting materials used. In Asia, a number of initiatives have been implemented for the production and cultivation of virus-free banana and citrus seedlings which have been integrated with other measures necessary for long-term control.

The development of sensitive serological and molecular diagnostic tools for these damaging viruses has improved greatly the impact of disease-free nursery schemes. In Taiwan, a national scheme for certification of citrus bud-wood has been underway since 1983 which employs an improved method of shoot-tip micro grafting, heat therapy combined with virus indexing that is conducted by the Plant Virology Lab (PVL) at the National Taiwan University (NTU). The stocks indexed as pathogen-free by the PVL/NTU are used as the pathogen-free citrus foundation stock collection for the program and are maintained in insectproof screen-houses. In Malaysia, several states have been involved in citrus rehabilitation programs and the use of pathogenfree planting materials has been recently implemented. The Malaysian Agricultural Research and Development Institute (MARDI) and Department of Agriculture, Sarawak maintain pathogenfree foundation stocks of popular citrus cultivars. Bud-woods from these plants are subsequently used for the propagation of pathogen-free planting materials. In Thailand, the disease-free Citrus Nursery Program was implemented by the Thai-German Integrated Pest Management (IPM) project in collaboration with the Department of Agriculture (DOA). The National Citrus Foundation Block was established in 1994. The DOA is responsible for the National Citrus Foundation Block and Certified Mother Tree Block. Disease-free bud-woods are produced from certified mother trees and released to approved citrus nurseries or farmer groups. A nursery accreditation scheme was introduced in 1993 in order to improve the quality control process.

Those nursery owners who wish to be involved in the scheme must follow the conditions, regulations and recommendations indicated in the scheme and must reapply for accreditation every 12 months. Nurseries are inspected twice a year. Similar programs also exist in Japan, Philippines (Phil-German Fruit Tree Project), Vietnam (Southern Fruit Research Institute [SOFRI] Citrus Disease-free Seedlings Program) and Indonesia (Indonesian Citrus Variety Improvement Program). In Taiwan, since 1983 a total of 26 million banana plantlets have been produced *in vitro* for commercial planting.

This has benefitted growers by making available pathogen-free plants and also in the rapid multiplication of new varieties. To meet the demand for large numbers of samples, virus-free stock foundations are kept in special facilities that are insect-proof and vector-free and the viral status is closely monitored.

These virus-free plants are used to produce mother plants that are used for multiplication by commercial nurseries. Meristem culture of banana for the production of pathogen-free plantlets has been developed elsewhere in the Asia-Pacific region. This approach is used in the Philippines as an ongoing eradication and banana rehabilitation program. A project for the control and rehabilitation of BBTV affected areas was implemented in 1994 and focused on a communitybased approach for smallholder banana growers. This involved consultations with officials and banana growers and a rapid rural appraisal (RRA) with farmer cooperators to determine their understanding of disease management practices.

This was followed by training on the early diagnosis of BBTV and proper eradication methods. Pathogen-free plantlets were distributed to farmers at four project sites. The results clearly indicate that areas previously devastated by BBTV can be made productive again. However, this can only be achieved with proper understanding of the disease by the farmers and appreciation of the need to integrate with other methods of control such as proper sanitation to remove alternative sources of inoculum and appropriate vector control.

Cultural Practices and Crop Disease Management

Many of the practices classified as cultural control have been practiced by farmers for centuries in the Asia-Pacific region and form a major component of their indigenous knowledge. Thurston has extensively documented many of these traditional practices and there are increasing demands that such practices and knowledge are integrated to ensure sustainable crop disease management. Traditional smallholder farmers in the tropics can rarely afford the cost of pesticides and labor is relatively inexpensive. This has meant that cultural control has and continues to play a major role in the management of crop diseases. In the Asia-Pacific region cultural control methods represent locality-specific options for disease management. They have proven their adaptability and worth in a particular environment. They have often arisen as a result of farmers' insightful knowledge of their farming systems and the pests within those systems. Dent has recognized the value of cultural control and believes that it should be the first line of defense in pest management programs. This can act as a foundation on which other control strategies can be implemented. Above all, cultural control is environmentally safe and is compatible with other control measures.

The following list of cultural methods has been shown to have some benefit for disease management in the Asia-Pacific region:

- *Destruction of Plant Disease Inoculum:* Many pathogens can survive in soil or on plant residues, and readily infect newly-planted crops. Such practices as clearing land and burning residues is believed to have a number of beneficial disease control effects. Soil-borne fungal pathogens are destroyed in the upper soil levels during burning. Flooding with water is a common feature of taro cultivation and may reduce pathogenic fungi in soil. The early removal (rouging) of virus-infected plants may help reduce the incidence of disease. In some countries, such as Hawaii and Guam, removal of watermelon mosaic virus infected plants has led to a reduction in disease incidence by providing less inoculum for insect vectors. The early removal of dasheen mosaic virus and *alomae-* and *bobone-*infected taros may provide good control. Banana bunchy top may be controlled in a similar way. It is important to be careful when removing infected plants as there is the risk that the vectors will disperse to uninfected plants. Plants may be doused in kerosene and burnt. Roguing of *Pythium*-infected taros may be useful and care should be taken to remove all decayed roots and corms. In Palau, the traditional approach is to remove the plant and all the soil surrounding the roots.
- *Crop Rotations:* Asian and Pacific farmers have used rotations for millennia. Particular plant pathogens are associated with specific crops. If these crops are continuously grown then these pathogens

can reach very high levels. By planting unrelated crops the buildup of disease is halted.

- *Plant Density and Spacing:* Low plant densities contribute to a reduction in humidity in the plant microclimate that lowers the incidence of a number of plant diseases such as Sigatoka of banana. In some Pacific countries increased plant spacing has been observed to reduce the incidence of TLB. In Asia, traditional spacing practices have been useful in managing downy mildew in maize. Sheath blight is now one of the major fungal diseases on highyielding varieties (HYVs) of rice in Asia. The HYVs tend to be planted in high densities which make them more susceptible to the disease. Prior to the release of IR8 in 1966 sheath blight was considered a minor rice disease. There is evidence that planting at low densities reduces the incidence of disease.
- *Depth of Sowing:* The perfect stands found in many crops grown by traditional farmers in Asia reflects their knowledge in using this practice.
- *Time of Planting:* Plant diseases can often be avoided by planting at times of the year that are unfavorable for disease development. In Asia, early planting of rice and maize has been reported to reduce the incidence of rice blast and downy mildew, respectively.
- *Flooding:* The paddy system of rice production represents one of the oldest practices that uses flooding for plant disease management. Severity of rice blast is reduced on flooded paddy rice compared to upland rice due to reduced infection periods. A similar system for taro production in the Pacific may contribute to disease management.
- *Mulching:* The placement of a mulch of old leaves or weeds around the base of beans has been shown to reduce infection by web blight. Mulching is believed to prevent the dissemination of soil-borne fungi by not allowing soil splashing by rain to occur. This has been demonstrated for yam anthracnose in the Pacific.
- *Raised Beds:* The use of raised beds for modification and cultivation of wetland areas has been described for a number of countries in Asia. In the Pacific, raised bed agriculture has been, and continues to be, effectively employed for the cultivation of mainly root crops. Raised bed systems have been described for a number of different environments and represent a highly productive form of cultivation. Thurston and Parker report that, in addition to their obvious irrigation and drainage value, there is little doubt that disease management is an added benefit of raised bed systems. They cite examples of the suppressive nature of raised bed soils against damping-off fungi (*Pythium* spp.) This is attributed to the increased biological activity of known *Pythium* spp. antagonists such as

Trichoderma spp., *Pseudomonas* spp., and *Fusarium* spp. in raised bed soils due to the addition of large amounts of organic matter. These antagonists are also likely to be active against other soil-borne plant pathogens. Raised bed soils were also found to suppress plant parasitic nematodes, containing nine organisms that had anti-nematodal activity. In addition, the high organic content of these soils was thought to contribute to the low numbers of nematodes in these soils.

- *Crop Diversity:* Polyculture or mixed cropping (the growing of two or more crops at the same time) has been observed to reduce the incidence of disease. It is believed that such environments represent a more stable ecosystem and offer a greater degree of protection compared with monocultures. A diversity of crops helps to delay the buildup of plant pathogens. It is interesting to note that the incidence of the disease kava dieback, caused by cucumber mosaic virus (CMV), is more severe in places such as Fiji where it is planted mainly as a monoculture. The incidence in Vanuatu where kava forms a component of the traditional mixed cropping system is lower. The virus is transmitted by an aphid vector and in the traditional mixed cropping system the aphid may have difficulty locating kava plants and along the way it will feed on other non-kava plants and quickly loses the ability to transmit the virus. There is abundant evidence to show that crop diversity contributes to the management of other pathogens also.
- *Destruction of Alternate Hosts:* Weed hosts act as carriers for a number of plant pathogens, especially viruses, over successive seasons. The control of these weeds is complicated, given that they are often symptomless carriers of viruses and root pathogens but their destruction has been demonstrated as contributing to a reduced incidence of certain diseases. Weed control is important for the control of a number of cucurbit viruses that cause damage in Pacific Islands and for control of CMV in bananas.

NONPESTICIDE CONTROL OF PLANT DISEASES

PHYSICAL CONTROL

Mulching and Screen Housing

There are two physical methods that are commonly used in Taiwan to prevent weeds, diseases or insect pests. For planting papaya, seedlings free from papaya ring spot virus (PRSV), the limiting factor of papaya production in Taiwan, are planted in screen houses that prevent aphids, the transmission vectors of PRSV, from entering. The soil is covered with plastic sheets that prevent the growing of weeds. By this way, papaya has been stably produced at about 138,600 mt per year since early 1990s. Mulching of soil with plastic sheets is also commonly adopted by the growers to cultivate many kinds of crops such as watermelon, muskmelon, cucumber, tomato, eggplant, pepper, strawberry, etc. By this way, weeds are mostly inhibited and the occurrence of insect pests and soil-borne diseases are also reduced to some extent. Besides, the simple facilities, tunnels constructed with transparent plastic sheets under which muskmelon is planted, are also commonly adopted by the growers. By this way, viral diseases, downy mildew and some insect pests are effectively controlled.

Steam Sterilization

Recently, plant disease control by steam sterilization was used in planting high-value horticultural crops in Taiwan. Soils and cultural composts contaminated with plant pathogens were sterilized by injecting steam (60-80 ºC for 30 minutes). Plant diseases, such as lily yellowing caused by *Fusarium oxysporium* Shlechtend:Fr. emend f. sp. *lilii* Imle, *Thanatephorus cucumeris* (Frank) Donk, bacterial wilt of lisianthus caused by *Ralstonia solanacearum* Yabuuchi, *et al.*, southern blight of ornamentals caused by *Athelia rolfsii* (Curzi) Tu and Kimbrough, etc., were controlled significantly by steam sterilization. In addition to the disease control, the quality of ornamental products and weeds control were also improved. Some horticulture growers, such as those who plant lisianthus and potted flowers, by the assistance of plant pathologists, developed their own steam pipe systems to improve the efficiency of operation in fields and screen houses. Although this method is labor-costly and time-consuming and, therefore, did not appeal to some growers, the benefits of reducing both the use of pesticides and adverse impacts on the environment still make it promising in the future, especially in the ornamental industry.

Table. Lethal Temperatures for soil Microbes and weeds

Lethal Temperatures (°C)	Soil microbes
100	All pathogenic microbes and weeds
93	Heat-tolerance viruses, actinomycetes and weeds
82	Most weeds and viruses, all plant pathogens
60-71	Most plant pathogenic fungi and bacteria, insects, mites earthworms, snals, centipedes

USE OF PATHOGEN-FREE PROPAGATING MATERIALS

Seeds and seedlings, the crop productivity and quality can be increased and improved. In Taiwan, the system of healthy seeds and seedlings production was started from the propagation and extension programs for seeds of the staple crops such as rice and coarse grains such as corn, soybean, peanut, etc. In 1959, the Seed Testing Laboratory (STL) was established for the inspection of fields for foundation, registered, and certified seeds production and laboratory testing of the quality including their purity, moisture content, germination rate and health. The acreage of fields, mainly paddy rice, inspected amounted to about 2,000 ha annually over the years. Subsequently, other propagation and inspection systems of virus-free seedlings of other crops were taken effect. Among them, the systems for potato, strawberry and sweet potato were established in 1972, 1991 and 1998, respectively. In the systems, STL are engaged in all the inspections and testing, and TARI, Chiayi Agricultural Experiment Branch Institute, and TSIPS are responsible for the supplies of foundation seeds of sweet potato, potato and strawberry, respectively.

The inspection systems of those three crops operated well and were successful in supplying high quality seeds and seedlings. The production technique of virus-free seeds of asparagus bean had been developed by TARI and worked in practice by TSIPS. The extension area of healthy seeds has amounted to 824 ha since 1996. With regard to passion fruit and bamboo, TARI and Tainan DAIS are responsible for, respectively, the inspection and maintenance of virus-free nuclear stock repositories founded to meet the needs of commercial nurseries. By estimate, 250,000 virus-free passion fruit seedlings have been released per year since 1989 and 60,000 healthy bamboo seedlings free from CMV have been provided since 1992. With regard to citrus, the system is carried out mainly by Chiayi Agricultural Experiment Institute with technical support from NTU. Pathogen-free budlings of citrus are provided by Chiayi Agricultural Experiment Institute to meet the needs of commercial nurseries. In recent years, over 20,000 pathogen-free budlings of citrus were produced annually. In addition, Taiwan Banana Research Institute and Taiwan Sugar Company take charge of the propagation and indexing of healthy seedlings of banana and sugarcane in Taiwan. The amount of healthy banana seedlings, produced by tissue culturing, supplied was estimated to 28 million during 1983-90.

Furthermore, to comply with the "Guidelines for Inspection of Designated Plant Diseases and Pests on Anthurium Seedlings", a mandatory inspection to ensure that the propagating materials are free from quarantine nematodes and bacterial blight pathogen, *Xanthomonas campestris* pv. *Dieffenbachiae* Vauterin, *et al.* has been implemented since September 1990 and over 690,000 healthy seedlings have been certified so far. Besides, the government is in the process of establishing of voluntary seed and seedling certification systems to encourage the production of high quality and healthy seeds. The certification system for oncidium orchid seedling will be the first to be taken effect and asparagus bean will be the second. In the future, many more crops will be included in the voluntary seed and seedling certification system when the techniques are available. It is expected that the implementation of the voluntary seed and seedling certification system will significantly enhance the quality of seeds and seedlings of related crops and subsequently promote the competitiveness of Taiwan's agriculture. The followings are detailed descriptions regarding some of the prominent cases that used pathogen-free propagating materials to control major diseases in Taiwan.

Application of Tissue Culture Technology

Fusarium wilt of Cavendish banana caused by *Fusarium oxysporium* Schlechtend:Fr. emend f. sp. *cubense* (E. F. Smith) Snyder and Hansen race 4 has been the major constraint for banana production in Taiwan. Previously, the disease had destroyed the banana industries of many countries in Central America despite many attempts to control the disease. Continued banana production in Taiwan in recent years to supply both local and the Japanese market was attributed to the successful development of tissue culture technology for controlling fusarium wilt. There have been two major achievements based on tissue culture technology leading to the success of controlling fusarium wilt. Firstly, the technology was employed for mass propagation of disease-free plantlets for commercial planting. During 1983-2001, the tissue culture program has produced a total of 28 million plantlets for planting over 14,459 ha and has effectively contained the spread of wilt pathogen via infected suckers traditionally used by farmers. Secondly, because propagation of banana by tissue culture has shown the potential of producing substantial genetic variability, a mass screening program to select tissue culture plants resistant to fusarium wilt was taken. Two resistant varieties, Tai-Chiao No. 1 and Tai-Chiao No. 3, developed by this breeding program were released for controlling fusarium wilt in 1992 and 2001, respectively. Another new, higher-yielding, resistant variety, Formosana, selected most recently is scheduled for release in 2002. It is estimated that planting Formosana to replace Giant Cavendish would have the potential to increase yield by over 50 per cent, and thus boost banana production considerably.

Plants of Green Bamboo

Because green bamboo shoot is rich in nutrition and delicious, it has been considered an important vegetable especially during the summer. BaMV is the most destructive disease affecting the cultivation of bamboo shoots. The percentage of infection on green bamboo based on visible symptoms was 80-98.8 in the major growing areas of green bamboo. Previous studies revealed that the disease was transmitted mainly by cultural practices via farming tools. The control for the disease by planting the virus-free clones was effective. Since 1989, eye-inspection and/or the ELISA (enzyme-linked immunosorbent assay) methods were used for the indexing of the disease, and green bamboo virus-free mother plant fields were set up at Tainan DAIS Hsinhua Branch Station in 1990. A virus-free clones nursery was set up at Paiho, and has been indexed periodically once a month and the new clones were distributed to the bamboo farmers at their request. This nursery has produced 60,000 virus-free clones for farmers until 2001. The establishment of the pathogen-free seedling system has significantly improved the quality and ensured the stable yield of green bamboo shoots. This system will also facilitate expanding the international markets of green bamboo shoots.

Control Virus Diseases of Asparagus Beans

Asparagus bean, *Vigna sesquipedalis* (L.) Fruwirth, is one of the important vegetables in Taiwan in hot and wet summer seasons. Among various kinds of diseases attacking asparagus bean, viral diseases are identified as major limiting factors for its production. CMV and BLCMV were previously found as the two major viruses attacking asparagus bean in fields. Both viruses were found seed-borne and the infected seedlings could serve as primary inoculum for the consequent early epidemic as resulted form mechanical and aphid transmission. Investigation was directed to evaluate the effects of planting virus-free asparagus bean seeds in controlling the viral disease in fields. An asparagus bean cultivar "San-Tse-Chin-Pi" was selected for production of virus-free seeds in 1990. By rouging virus-infected plant after detection by ELISA using antisera against BLCMV and CMV, plants were cultivated in insect-proof screen houses until seeds were harvested. The seeds were then served as foundation virus-free stock. A large-scale virus-free seed propagation system was subsequently developed in order to satisfy sufficiently the demand from the growers in Taiwan. In 1991-92 we conducted four field trials to test the feasibility of using virus-free seeds to control viral diseases of asparagus bean. The result indicated that the application not only significantly delayed the virus epidemic but also increased the yield and quality of bean pods and eventually the profit earned by the growers. In the field experiments, as low as 30-72 per cent of final virus incidences was recorded in the plots sowed with virus-free seeds while in the same time 100 per cent of incidences were detected in the control plots planted with commercial seed lots. Since 1993, planting of virus-free seeds has been routinely taken as one of the key measures in the integrated pest control program of asparagus bean in various production areas. Because of the consistent and significant control effect the virus-free asparagus bean seeds, produced in this program, have become more and more popular in Taiwan with time. Over 1,086 kg of virus-free asparagus bean seeds were produced and released to four different major asparagus bean producing counties in 2001.

MAJOR VIRUS DISEASES IN GREENHOUSE CROPS

Table summarizes the characteristics of the main virus species that cause diseases in protected crops, for which comprehensive reviews are available. Some of these species have been further reviewed in the text.

Aphid-Transmitted Viruses

Cucumber Mosaic Virus (CMV)

Description. CMV is the type species of the genus Cucumovirus of the family Bromoviridae of plant viruses. CMV virions are 29 nm icosahedrical particles that encapsidate a single-stranded RNA genome of messenger sense divided in three molecules, RNA 1, 2 and 3 (1.3×10^6, 1.1×10^6, and 0.8×10^6 daltons, respectively).

Some CMV isolates encapsidate an additional small RNA called satellite RNA (0.1×10^6) daltons), that depends on virus for replication, encapsidation and movement. RNA satellites are able to modulate the symptoms induced by CMV. A great variability among CMV isolates has been reported. According to biological properties of symptomatology, thermosensitivity *in vivo*, molecular and serological characteristics, most CMV isolates have been assigned to two main groups.

Transmission, Host Range and Diseases

In nature, CMV is transmitted in a non-persistent manner by more than 60 aphid species including *Aphis gossypii* Glover, *Macrosiphum euphorbiae* (Thomas) and *Myzus persicae* (Sulzer). Variable rates of seed transmission have been described in 20 species including some vegetable crops like bean or spinach, or weeds like *Stellaria media* Cyrill. There is no evidence of seed transmission in cucurbits. CMV can be mechanically transmitted in experimental conditions.

CMV has an extremely wide host range that comprises more than 1000 species of dycotyledons and monocotyledons. Host range includes many important vegetable crops like melon, cucumber, zucchini squash, watermelon, tomato, pepper, eggplant, lettuce, carrot, celery, spinach, pea, etc.; ornamentals like anemone, aster, dahlia, delphinium, geranium, lily, periwinkle, primula, petunia, viola, zinnia, etc.; and woody and semiwoody plants like banana, ixora, passion fruit, etc. Symptoms are extremely variable depending on the CMV isolate, host species or cultivar, plant age at infection time, and environmental conditions. Early infected plants can show marked stunting. Symptoms in leaves are mosaic, mottle and/or distortion. Necrosis is induced by certain isolates. Flower abortion and fruit discoloration and malformations are caused.

Economic Importance and Control

CMV is distributed worldwide, predominantly in temperate regions but with increasing importance in tropical countries. It causes serious diseases in many important crops grown in the open but also in protected conditions (tomato, pepper, cucurbits, etc.). Yield reductions are mainly due to decreased fruit set, and production of non-marketable fruits because reduced size, or presence of symptoms like mosaics, malformations or necrosis. Control of CMV is difficult because of the wide host range and its rapid natural transmission by aphids.

Integrated control measures are recommended in protected crops to reduce CMV incidence:

- Elimination of infected plants;
- Avoidance of aphid entrance in the greenhouse by covering entrances with aphid-proof nets;
- Reduction of aphid populations by using insecticides;
- Use of virus-free seeds (for example in bean and spinach);
- Elimination of alternative spontaneous hosts present in and around the crop.

Resistance to CMV is available in cucumber and programmes are in course in melon using Korean and Chinese varieties. Sources of resistance or tolerance have been found in most cultivated or related species. However, in most cases resistance or tolerance is not absolute, and is overcome by some CMV species. Aphid vector tolerance or resistance incorporated in the plant can be combined with other control methods. Transgenic melon, cucumber and squash plants expressing the coat protein gene of CMV offer a good level of resistance to several strains of the virus.

Description

The Potyvirus genus of the family Potyviridae is by far the largest of the plant virus groups. Many members cause important economic losses in protected crops and can be a major limiting factor for production. Virus particles are elongated and flexuous (680–900 × 11 nm) with one molecule of messenger sense single-stranded RNA (3.0-3.5 × 10^6 daltons) attached covalently to a protein. The genomic RNA codes for a large polyprotein that is proteolytically cleaved to yield the mature viral proteins. Virus infections are associated with characteristic cytoplasmic and nuclear inclusions, pinwheels, bundles and laminated aggregates.

Transmission, Host Range and Diseases

Potyviruses are transmitted in nature by aphids in a non-persistent manner. Some aphid species (especially those of the genera *Myzus, Aphis* and *Macrosiphum*) are associated with high virus incidences in crops. Seed transmission is important epidemiologically in certain potyviruses, like bean common mosaic virus (BCMV) in French bean, or LMV in lettuce. Potyviruses can be transmitted experimentally by mechanical inoculation. In nature, most potyviruses have relatively narrow host ranges, few species within one genus or closely related genera; for example: BCMV is restricted to *Phaseolus* species; potato virus Y (PVY) to members of the Solanaceae; watermelon mosaic virus2 (WMV2), zucchini yellow mosaic virus (ZYMV) and zucchini yellow fleck virus (ZYFV) mainly to species of the Cucurbitaceae; and LMV to species mainly in Compositae. Potyviruses can induce severe diseases in important crops. Symptoms may vary depending on host species, virus strain, environmental conditions and plant age at infection time. Potyviruses like ZYMV, WMV2 and papaya ringspot virus-W strain (PRSV-W) can cause severe diseases in zucchini, squash, melon, cucumber and watermelon, inducing stunting, chlorosis, mosaic, leaf malformation, flower abortion, and fruit and seed malformation. Vein clearing, mosaic, yellow mottling and growth reduction are often observed in LMV infections of lettuce, endive and spinach. Legume infecting potyviruses like BCMV cause abnormal formation of seeds that are smaller, discoloured and/or distorted.

Economic Importance and Control

The Potyvirus genus is the most devastating among plant viruses. Damaging members like BCMV, bean yellow mosaic virus (BYMV), ZYMV, WMV2, PRSV-W and PVY are spread worldwide and cause economically important problems where present. Several authors reported losses up to 100 per cent in squash, cucumber and watermelon caused by ZYMV. Potyviruses are mainly a problem in outdoor crops, however, can also be a severe threat in protected crops.

Control should be done by an adequate management of the crop, integrating different control measures: if seed-transmitted, the use of certified virus-free seeds is the basis for effective control; use of virus-free plantlets will avoid primary infections; because transmitted in a non-persistent manner, spraying insecticides is not effective for preventing virus spread, however, effective control has been obtained in some cases by spraying with light mineral oils in outdoor crops.

Successful breeding programmes for resistant cultivars have been done in lettuce to LMV (resistance breaking strains have recently been described), in French bean to BCMV, and in melon to PRSV-W. Transgenic approaches have also been explored, overcoming difficulties associated with conventional breeding methods. Cross protection using an attenuate poorly aphidtransmissible strain of ZYMV (ZYMV WK) have been successfully used to control ZYMV in cucumber, melon and squash.

Luteovirus Genus Description, Transmission, Host Range, Diseases and Economic Importance

There are a number of yellowing diseases transmitted in nature by aphids that are caused by viruses in the Luteovirus genus. This is the case of beet western yellows (BWYV) and cucurbit aphid-borne yellows (CABYV) viruses. Viral particles are 25–30 nm icosahedral, and encapsidate a monopartite, single stranded, messenger sense RNA genome. Transmission in nature is by aphids in a circulative, non-propagative, persistent manner.

BWYV infects lettuce, cucumber, watermelon, squash, sugarbeet, carrot, spinach, pepper and tomato, symptoms being mild chlorotic spotting, yellowing, thickening and brittleness of older leaves. It has been reported in North America, Europe and Asia, and is probably distributed worldwide. CABYV causes a yellowing disease of melon, cucumber and zucchini squash; symptoms are initial chlorotic patches, leaf thickening and general bright yellowing of leaves.

In melon and cucumber important yield losses are reported, due to reduced number of fruits per plant caused by flower abortion but not by altering fruit shape or quality. It was first described in France in outdoor and protected crops and has been found through the Mediterranean area, Asia, Africa and California.

Control

Disease management should be by integrating measures to reduce aphid populations within the greenhouse via avoidance of insect entrance (nets in windows) and chemical spraying, with measures to reduce infection foci (virus-free planting material, elimination of infected plants). Sources of resistance have been found for CABYV in melon germplasm, and for BWYV in lettuce.

SPATIAL DISTRIBUTION OF DISEASED UNITS IN A PATHOSYSTEM

Spatial distribution of diseased units in a pathosystem is the most important factor affecting the field estimation of disease intensity. Spatial distribution includes the way in which disease lesions are distributed among healthy units and the way in which diseased host units are distributed among healthy units. Distribution of diseased units may be random, aggregated or regular. With randomly distributed disease, the variance is theoretically equal to the mean. In aggregated patterns, the variance in the number of lesions per leaf is greater than the mean number of lesions per leaf, but when there is a regular pattern the variance is smaller than the mean. When a large number of host units are sampled for disease, a frequency distribution showing the number of diseased units in each severity category may be determined. The sample frequency distribution can be compared with theoretical distributions using the goodness-of-fit test, and the parameters of the empirical distribution may be defined.

Theoretical distributions applied to biological systems include the normal, log normal, Poisson, Weibull, Gamma and negative binomial ones. Knowledge of the frequency distribution is essential for the design of sound sampling procedures. When estimating disease intensity per field, the sampling unit, sample size, sampling point, sampling fraction and sampling method must be considered. In most disease assessments, the sampling unit is a plant. Often, only selected parts (such as individual leaves) may be assessed for disease intensity.

For each field, a predetermined number of sampling units is selected to give a mean value representative of that field; this is the sample size. Sample size is determined by the cost of sampling, the precision required and the available time, as well as by the spatial distribution of the disease; that is, the sample size should be empirically defined. Many sampling methods have been reported for plant disease assessments. Samples may be taken at intervals along predetermined lines in the field or greenhouse and these may be either one diagonal, both diagonals (forming a big letter X), or (if a more representative sample is required) a large W or Z pattern. With a disease that is randomly distributed, all the above methods will give comparable results and reduced variance in the sample mean may be better achieved by increasing sample size. If the diseased units are aggregated, the sampling method will be more important than sample size, and the large X or W sampling pattern is preferable to the single diagonal.

MONITORING PATHOGEN POPULATIONS

Monitoring the pathogen, primarily by trapping air-borne spores, has been used as a measurement of disease intensity and development and could serve either as an alternative, or a complement, to disease assessment. Given the current technology, the use of spore counts of pathogen populations for field measurement of disease is unlikely to replace the main conventional methods of measuring disease severity (disease symptoms), unless its accuracy can be shown to override the ease and low cost of symptom assessment. The monitoring of pathogen populations may serve another purpose. As fungicides still remain an important tool for control of plant pathogens in the greenhouse, it is important to monitor populations of the pathogen for their resistance to potential fungicides. The term "monitoring resistance" is used to denote testing for sensitivity of target organisms in field populations. This can range in scope from continuous surveillance programmes over several years and involving many locations to short-term investigations into individual cases of suspected resistance. Good monitoring is the cornerstone of fungicide resistance research.

Without such work, we would know virtually nothing about the occurrence of resistance in crop pathogens. Moreover, resistance monitoring, together with monitoring for changes in practical performance, is a vital component of integrated resistance management. Several tools have been developed for such a purpose. For example, a tool for estimating the resistance of populations of *Botrytis cinerea* Pers.:Fr. to common fungicides has recently been developed. Tested fungicides are added to a selective medium in Petri dishes. The plates are exposed in the greenhouse at approximately midday, when *B. cinerea* conidia are released into the air. Plates are exposed for 30–60 min, according to the intensity of the disease in the greenhouse and then incubated for 4–7 days. Counts of typical *B.*

cinerea colonies in the media supplemented with the fungicides are compared with those from fungicidefree plates. The data may then be used to make a recommendation on fungicide use.

Concluding Remarks

Studies on population dynamics of insect pests or beneficials and plant pathogens, which have been performed to improve the efficiency of IPM, have followed parallel paths and run into similar obstacles. Because of the speed of the dynamics and the strong spatial heterogeneities of these populations, control strategies have had to be designed to include the large amounts of data that may be generated over different temporal and spatial scales. In both disciplines, methods have been designed to evaluate quickly insect densities or levels of disease injury in large and frequent samples. Moreover, the need to sample commercial crops to take into account large-scale variations implies the use of non-destructive methods. Pathologists and entomologists have independently concluded that visual indices could be practical and efficient. Initially, both of these groups have tried the two-class (presence/absence) indices, and then later the several-class indices.

After it was discovered that the logarithmic scale is a natural tool of the human eye discriminating among different kinds of intensities, statistical approaches were developed to evaluate the precision of such evaluations. In the future, many new methods will need to be constructed to advance IPM strategies. A good idea would be to synchronize some of these developments in the two disciplines (*i.e.* for all the major insects and pathogens of a given protected crop), and to pool the statistical approaches which, as a matter of fact, deal with very similar problems. Such an integration would allow the pathologists and entomologists to propose standardized "toolboxes" to professional and technical partners.

CONTROL OF PLANT DISEASES WITH ANTAGONISTIC BACTERIA

Bacillus subtilis Cohn, a gram-positive endospore-forming bacterium ubiquitous in soil and many plant surfaces, has long been used as a beneficial microbial agent for the control of plant diseases. The fungicidal application of *B. subtilis* was one of the primed focuses of biopesticide product development during the past decade and attempt of commercialization development was increasing. In Taiwan, a *B. subtilis* isolate was recently commercialized by Kuan-Hwa Chemical Co. The product has been registered for the control of powdery mildew of pea, whereas its potential market remains to be evaluated. In order to explore the usefulness of native bacilli resources, *Bacillus* spp. with broad spectrum of antagonistic activity against various phytopathogenic fungi and bacteria were selectively isolated from cultural substrate and soil samples collected from field. Among them, isolate BS1 of

B. subtilis was found outstanding because of the superior antagonistic activity against *Rhizoctonia solani* Kühn, *A. rolfsii* and *Xanthomonas campestris* pv. *mangiferaeindicae* Robbs, *et al.* Its mass production techniques were developed.

As for the practical uses, more than 30 crop species have been tested in greenhouse and/or field trials. Effective control of soil-borne and foliar diseases has been demonstrated, and significant growth promoting effect was generally observed. The efficacy of disease control apparently had certain connection to the production of iturin-like antibiotic.

By spraying application of the liquid formulation at 500-1000 x dilution, effective control of foliar infection of tobacco by *Cercospora* sp. (tobacco white spot disease), and leaf and fruit infection of mango by *X. campestris* pv. *mangiferaeindicae* (bacterial spot of mango) were demonstrated by field trials. In cooperation with private enterprises, the commercialization of BS1 isolate is in progress. *Streptomyces* spp., being gram positive also, are found ubiquitous among various terrestrial and aquatic natural habitats. The members of streptomyces mostly are known to be saprophytic, aerobic, exist commonly in soil, grow well in slightly alkaline environment, and adapt well in high temperature and drought conditions.

As many *Streptomyces* spp. are known to produce important antibiotic secondary metabolites and hydrolytic enzymes, the wide and extensive industrial application has been well recognized. In Taiwan, a *Streptomyces saraceticus* Berger, *et al.* isolate, SS31, was selected and was shown to be effective in knocking down rootknot nematode population in soil primarily because of its chitinase producing characteristics. In a preliminary screening test, it was noted that the test isolate was strongly antagonistic against fairly wide spectrum soilborne fungal pathogens – especially members of Oomycetes. When solid formulation of SS31 was added to a soil amendment consisting mainly spent forest mushroom compost, fish meal, crab shell powder, and rapeseed pomace, and amended into the soil, the mixture effectively controlled fusarium wilt of lettuce, and fusarium wilt of banana. By spraying application of the liquid formulation at 1000 x dilution, effective control of Phytophthora diseases of papaya and citrus has been demonstrated. It was also found effective for the control of downy mildews on cucurbit, rose, grapevine, etc. In addition, the formulation was found useful as soil drenching agent for the control of soil-borne diseases. The drenching treatment of the liquid formulation at 100-300x in dilution significantly reduced damping-off of cucumber seedlings caused by *Pythium aphanidermatum* (Edson) Fitz. In cooperation with private enterprises, the commercialization of SS31 isolate is in progress.

CONTROL OF PLANT DISEASES WITH ANTAGONISTIC FUNGI

In Taiwan, *Trichoderma viride* Persoon:Fries and *Trichoderma koningii* Oudemans had been reported to be pathogenic to mushroom in 1967 and water chestnut in 1955, respectively. However, most of species/ strains of *Trichoderma* were found to be the effective microorganisms for suppression of the plant pathogenic fungi after 1980. During 1980-90, antagonistic *Trichoderma* spp. were studied and the promising ones were selected, formulated and added into soils and found to be effective on the control of several plant diseases including *Rhizoctonia* diseases, *Sclerotinia* disease, *Pythium* damping-off and *Phytophthora* blight. Among them, *T. viride* T20, *T. knoningii* T12, *T. knoningii* T12-R33-D25, and *Trichoderma harzianum* Rifai TVCN2 were tested in large-scale trials. The results showed that they could inhibit *Rhizoctonia* disease of chrysanthemum and adzuki-bean. Recently, it was found that *T. virens* strain R42 and *T. harzianum* T22, when added to the soils, could induce cucumber resistance to cucumber green mottle mosaic virus and suppress the sudden death of melon caused by *Monosporascus* spp. Both of the biological control agents also promoted the seedling and root growth of the crops in greenhouse and field trials. In addition, two isolates, *T. harzianum* T2 and *T. virens* T3 that had been formulated as granules, were found to be effective on the control of dollar spot of turf grasses caused by *Sclerotinia homoeocarpa* Bennett, when the dose applied had reached 10^6 cfu/g soil dry weight. The techniques for the mass production of the antagonistic fungi have been exploited.

USE OF SOIL AMENDMENTS

Eight formulated soil amendments, S-H mixture, SF-21 mixture, L-T mixture, AR-3 mixture, CH-1 mixture, GS mixture, FBN-5A and CF-5 liquid, have been developed, mostly by plant pathologists at NCHU, and used for the control of soil-borne plant pathogens in Taiwan. The ingredients of these compounds are mainly fertilizers and organic wastes. Among them, S-H mixture and L-T mixture have been commonly used and FBN-5A is ready for commercialization.

Control of Fungal Diseases with S-H Mixture

S-H mixture, developed in 1983, has been commercialized in Taiwan for the control of soil-borne diseases. Amendment of soil with S-H mixture, significantly reduced the incidence of watermelon wilt caused by *Fusarium oxysporium* Schlechtend:Fr. emend Snyder and Hansen f. sp. *niveum* (E. F. Smith) Snyder and Hansen, radish yellows caused by *Fusarium oxysporium* Schlechtend:Fr. emend f. sp. *raphani* Kendric and Snyder, pea wilt caused by *Fusarium oxysporium* Schlechtend:Fr. emend f. sp. *pisi* (van Hall) Snyder and Hansen, fusarium wilt of banana caused by *F. oxysporium* f. sp. *cubense,* clubroot of crucifers caused by *Plasmodiophora brassicae* Woronin, cucumber damping-off caused by *P. aphanidermatum,* southern blight of pepper caused by *A. rolfsii, Rhizoctonia* blight of bean caused by *R. solani* and bacterial wilt of tomato caused by *R. solanacearum.* The effective rate of application for the effective rate of application for S-H mixture varied with diseases and crops, ranging from 0.5 to 2 (w/w) of soil. In more than 5,000 ha of field trials in a variety of vegetable crops, S-H mixture not only reduced disease incidence but also improved root health, plant growth and plant stands.

MANAGING THE CROP ENVIRONMENT

TEMPERATURE

In very general terms, diseases as well as arthropods can be said to have optimum temperatures for their dispersal and development but these cardinal points are the integral of the optima of several growth phases of the pathogen as well as of different defence reactions of the host Jarvis cited different temperature optima for different growth processes in the grey mould pathogen *B. cinerea:* mycelium growth, sporulation, conidium germination, germ tube growth, appressorium formation, sclerotium formation and sclerotium germination. All have different temperature optima, most of which lie above the general optimum range for grey mould development, 15–20 °C. In most of its hundreds of hosts, resistance to *B. cinerea* is probably least within that range. The temperatures of leaves and fruit can vary markedly from ambient air temperatures as determined by conventional greenhouse instruments, and so the temperature within the boundary layer can be assumed also to be different.

At night, energy lost by radiation from leaves can result in temperatures 1–3 °C cooler than ambient air and temperatures frequently reach the dew point. In crops transpiring well, evaporative cooling can also reduce leaf temperature but insolated leaves not transpiring can become considerably warmer, by as much as 2–8 °C, than ambient air. Similarly, Schroeder found that the temperature of red tomato fruits rose from about 20 to over 50 °C in air that rose from 26 to 37 °C in the same period. On the other hand, green fruits exposed to the same conditions remained 4–8 °C cooler than the red ones. Temperatures of leaves, flowers and fruit can be considerably decreased by shading from direct sun and by increasing evaporative cooling by adequate ventilation and forced air flow. Eden *et al.* discussed the possibilities of raising flower truss temperatures in tomato crops to avoid grey mould. Whereas higher temperatures resulted in increased numbers of flowers infected by *B. cinerea,* the fungus was less likely to grow proximally to the main stem where the damage would be far more severe than one infected flower. On the other hand, higher temperatures (20–25 °C) resulted in fewer infections of stem wounds than at 15 °C.

Eden interpreted these results in terms of changing balances between fungal aggression and host defence reactions. Just as with diseases of shoots, temperatures can be to some extent selected to minimize diseases of roots; for example corky root rot *(P. lycopersisci)* of tomato can be largely avoided by transplanting into warm media at 20 °C, as can Fusarium crown and root rot (F. *oxysporum* f. sp. *radicis-lycopersici).* By contrast, the optimum temperature for the expression of Fusarium wilt is 27 °C. Similarly, *Pythium aphanidermatum* (Edson) Fitzp. is most pathogenic to spinach in hydroponic culture at 27 °C, whereas *Pythium dissotocum* Drechs. is most pathogenic at 17–22 °C. It is therefore important to know exactly which of closely related pathogens is present. Insects and mites, like diseases, have also an optimum temperature for their activity, dispersal and development Generally, greenhouse pests are thermophilic and perform best within 20–30 °C night-day ambient temperatures.

The preferred temperature for aphids and the greenhouse whitefly is somewhat lower, 15–25 °C. The interaction between temperature and VPD on the survival of western flower thrips was determined by Shipp and Gillespie. Of course, temperature affects not only arthropod pests but also their natural enemies. Natural enemies may perform poorly if temperatures are too high or too low which may occur during summer and winter respectively in the Mediterranean area. Then, the more temperature-tolerant *Diglyphus isaea* (Walker) or *Dacnusa sibirica* Telenga can be used according to thermal regimes expected in greenhouses. Excessive heat, combined with high VPD is a serious constraint for *Phytoseiulus persimilis* Athias-Henriot in warmer Mediterranean areas. Shipp and van Houten determined optimum temperatures and VPD for the use of *N. cucumeris* in Canadian cucumber houses, and these types of studies serve as guides to more intelligent biological control.

Humidity

The effects of humidity on greenhouse crops have been reviewed by Grange and Hand, and their direct and indirect effects on diseases by Jarvis. Uncertainty about VPD and temperatures in the boundary layer raises considerable suspicion about the validity of countless experiments on the infective abilities of fungal spores and disease prediction systems at low VPDs and inadequately measured or inadequately controlled temperatures. Fungal spores and bacteria require a wet substrate in which to initiate infection, and the water on leaves and fruits is provided by dew, guttation or overhead irrigation. This last can be discounted in well-managed greenhouses as an invitation to pathogens. Fogging systems cooling the air by evaporation are permissible if all the droplets evaporate before they land on plants. Measuring the onset and disappearance of dew is very difficult without the sensors themselves altering the boundary layer microclimate by heat conduction, shading, etc. Wei however, developed a copper-coated polyamide film sensor that could be wrapped around a tomato fruit and which had a response time of only a few seconds from dry to wet, and a response of less than 2 minutes to Peltier cooling of the surface to dewpoint.

Connected to microclimate modifiers (heating, ventilation), this device could obviate much of the risk of infection. Predicting the onset of condensation and its evaporation is even more difficult using atmospheric variables such as relative humidity, temperature, airspeed and radiation. Most predictions have errors in excess of 0.8 h and as much as ±3h. Clearly this is unacceptable in a cucumber house where infection of flowers by *Didymella bryoniae* (Auersw.) Rehm can occur in 1–2 h. Modelling the duration of dew in situations other than greenhouses has been done but with wide differences between predicted and observed durations of wetness. When the dewpoint temperature of the air falls below the temperature of the plants in a greenhouse, they become covered with water droplets and films, perhaps with hydrophilic fungal spores as nuclei, especially in still air at low VPD. Wei developed a model from heat transfer theory that accurately simulated condensation and evaporation from tomato fruits still attached to the plant:

$$E_L = \frac{\rho\alpha Cp\left[e_a - e_s(T)\right]}{r_d\gamma}$$

where E_L is the latent heat flux, $\tilde{n}_a$ is the density of air, *CP* is the specific heat of air, *e,* and e_s (T) are vapour pressures of air and saturated vapour pressure of air, respectively at T°, r_d is the boundary layer resistance to vapour transfer between the wet surface and the air, and g is a psychometric constant. Using the wetness sensor of Wei obtained excellent agreement between simulated and measured fruit surface temperatures during condensation and evaporation, within 0.3–0.5 °C (standard deviation 0.4 °C). The model predicted wetness within 5 minutes of its detection, and dryness came as predicted. Clearly, this precision gives ample time for preventive action to be taken against most fungal infections. While free water and low VPD are to be avoided if pathogens are present, those very conditions are needed to establish epidemics of fungal pathogens of insects, such as *Verticillium lecanii* (A. Zimmerm.) Viégas, *Beauveria bassiana* (Balsamo) Vuillemin and *Paecilomyces fumosoroseus* (Wize) Brown and Smith. Similar contrary indications have been obtained for arthropod pests and their predators. While spider mites are most active at relatively high temperatures and low VPDs, their predator *P. persimilis* is inhibited in those same conditions. Optimum humidity conditions for the predatory activity of *N. cucumeris* has been established by Shipp and van Houten.

Water Stress

Guttation results when the rate of water supply osmotically pumped by the roots exceeds the rate of water lost by transpiration and used in growth. To prevent guttation, the osmotic potential of the root xylem must be more negative than that of the nutrient solution. In poorly-managed greenhouses, guttation frequently happens at night when VPDs are low and root temperatures maintain high metabolic activity and root pressure. Tissues become waterlogged (oedema) and water guttates from stomata and from hydathodes at leaf margins with profound effects on the phylloplane microorganisms. Water continuous with the surface and substomatal vesicles facilitates the entry of bacteria into leaves of for example *Pelargonium* spp., particularly when resumed transpiration leads to resorption of the water. Wilson described how reversal of transpiration flow permits conidia of *B. cinerea* to enter tomato stem xylem, there to remain a latent inoculum. Water alternately accumulating and evaporating from hydrothodes leaves toxic deposits of salts, a ready entry point for necrotrophic pathogens. Lesions of gummy stem blight (*D. bryoniae*) are frequently seen originating from such points on cucumber leaves. Guttation damage can easily be eliminated by regulating atmospheric humidity, ventilating effectively, reducing evening watering and adjusting the osmoticum of nutrient solutions.

Light

Setting aside the effects of daylength on flowering in florists' crops, photosynthetically active radiation (PAR) (400–700 nm) is the part of the spectrum with the greatest effect on crop growth and productivity. Low and high light intensities are important agents of stress in crops that induce physiologic strains predisposing the crops to disease.

Particularly important is the effect of light combined with crop management procedures, such as plant spacing, row orientation, training and pruning systems, irrigation and nutrition, on the partition of assimilates, and the relative susceptibility of different tissues and organs to disease. Daylength, however, is important in determining diapause in both arthropod pests and their predators. Early diapause may be a major constraint in their use. Non-diapausing strains can, to some extent, overcome this problem. Light also has direct effects on fungal sporulation, germination and sclerotium formation. In *B. cinerea,* most isolates are stimulated to form conidia by light in the near-UV band (320–380 nm), an effect temporarily reversed by blue light. Some isolates, however, form conidia in the dark.

All fungi grow mycelium in the dark, and *B. cinerea* forms its sclerotia in darkness, or in yellow or red light, or when irradiated for less than 30 min with near-UV light. The requirement of B. *cinerea* and some other fungi for near-UV light for sporulation has led to the development of greenhouse covering materials that screen out that band as a means of disease control. Tuller and Peterson found fibreglass to transmit much less light of 315–400 nm than did polyethylene but in a comparative assessment of grey mould in Douglas fir seedlings [*Pseudotsuga menziesii* (Mirb.) Franco] it was concluded that the principal effect of low irradiance transmitted by fibreglass was in inducing needle senescence in dense canopies and thus susceptibility to grey mould, rather than on any direct effect on fungal sporulation. In both types of greenhouse, the mean intensity of light that inhibited sporulation (430–490 nm) exceeded that that promoted sporulation (300– 420 nm). In those greenhouses, too, predisposing conditions of temperature (15–20 °C) and humidity (>90 per cent RH) persisted 14.5 times longer in fibreglass than in polyethylenecovered houses.

Humidity effects also seem to have outweighed effects of light wavelength in a series of trials with coloured cloches covering strawberries. Grey mould was most severe under pink and blue plastic covers, where VPDs were lower (0.41 and 0.64 kPa, respectively) than under clear plastic (1.14 kPa), or under glass (1.74 kPa). The effects of light are evidently not simple. Nevertheless, attempts have been made to filter out the near-UV light that induces sporulation in some fungi. Reuveni incorporated hydroxybenzophenone into polyethylene, which increased the ratio of inhibitory blue light (480 nm) to UV (310 nm), and reduced the sporulation of *B. cinerea* in polystyrene petri dishes.

Under the treated plastic, grey mould lesions were fewer in tomato and cucumber (17 and 15, respectively) than under untreated plastic (41 and 29, respectively). Similarly, plastic coverings absorbing light at 340 nm inhibited the sporulation and reduced the incidence of grey mould lesions on cucumber and tomato as well as white mould lesions caused by *Scerotinia sclerotiorum* (Lib.) de Bary. Many isolates of *Altenaria solani* Sorauer also depend on near-UV light for sporulation, and Vakalounakis used vinyl films filtering out light of <385 nm to reduce the incidence of early blight in tomato greenhouses to less than 50 per cent of that under unamended vinyl film. Except as an agent of stress on the host, light has little direct effect on the rhizosphere microflora.

Carbon Dioxide and Oxygen

Carbon dioxide enrichment is a standard procedure in many commercial greenhouses but because it necessarily involves some restriction in ventilation to achieve the concentrations of CO_2 required, of the order of 1000 vpm, there is often increased danger of unmanageable low VPD. The concentrations of CO_2 that impair the growth of *B. cinerea* are 2–3 orders of magnitude greater than those found even in CO_2-enriched greenhouses and so reports, for example, of Winspear, of increased incidences of grey mould in CO_2-enriched greenhouses, can be interpreted in terms of enhanced levels of assimilates, or a denser canopy, with its increased risks of disease-susceptible wet plants. While CO_2 is a prominent component of the rhizosphere atmosphere as a product of root and microbial respiration, it has little direct effect on pathogens.

Oxygen deficiency stress readily occurs in compacted and waterlogged soils and in over-warm hydroponic solutions in which both increasing temperature and increasing solute concentration decrease oxygen solubility. Further, increased temperatures lead to higher root and microbial respiration rates which further deplete oxygen tensions. Low oxygen tension has been advanced as an explanation for physiological root death as well as decreased host resistance to root pathogens.

Air Movement

The primary purposes of directing and regulating air movement in the greenhouse are:

- To reduce the steepness of gradients in temperature, vapour pressure deficits and CO_2;
- To assist in the evaporation of infection droplets;
- To induce thigmomorphogenesis in bench-grown crops. This last results in sturdier plants and resistance to Fusarium wilts.

Through-the-bench air movement and plant spacing on the bench are important factors in escape of forest seedlings and *Exacum affine* I.B. Balf. ex Regel from grey mould. Counter to the generally beneficial effects of air movement are its effects on pathogen spore dispersal. Most fungi sporulate best in still air at VPD of 1.2–0.6 kPa but fungi of the Peronosporales, like *Bremia lactucae* Regel and *Pseudoperonospora cubensis* Rostovzev sporulate on wet surfaces. Airborne conidia are often liberated from conidiophores by hygroscopic mechanisms and are dispersed by air currents. Both mechanisms rely on disturbance of the microenvironment such as is readily provided by worker activity. The same mechanisms that control the liberation and dispersal of pathogen spores also apply to spores of biocontrol fungi when control is by enhancement of indigenous populations. Air movement also effects the passive transport of spider mites on webs floating through the air and being trapped on neighbouring plants. Forced air flows can transport larger insects into the greenhouse, even through some screens. Aggregation of insects is controlled by airborne semiochemicals, while the dispersal of pheromones on excessive air currents can interfere with mating disruption as a means of biological control, or attraction into sticky traps.

Integration of Environmental Factors

Epidemics of diseases are the result of a complex sequence of biological events each with a different set of permissive environments that have to occur in sequence, and coupled with hosts in a receptive state. Jarvis outlined the complexity of those events in the case of grey mould epidemics. Beginning with sporulation, conidia are formed at temperatures around 15 °C and in moderate VPD; they are liberated by hygroscopic movements of the conidiophore in rapidly changing conditions of humidity, and are dispersed on air currents or by water-splash; infection occurs on wet surfaces at 15–20 °C; and colonization of the host is fastest at 25–30 °C.

Marois found that epidemics of grey mould on rose depend as well on inoculum concentration, a relationship that was different in winter and summer, and affected by temperature, relative humidity and VPD, the latter the far more meaningful parameter for describing epidemiology of *B. cinerea* in roses. It has been possible to construct working models of grey mould epidemics in cucumber; tomatoes; gerbera and rose; and conifer seedlings.

The value of epidemic models such as BOTMAN, an integrated chemical and biological control programme, in predicting the onset and course of epidemics, however, is severely compromised by the rapidity with which infection occurs – 9–10 h for grey mould and only 1 h for gummy stem blight in cucumber flowers – and by the wide variability of the greenhouse climate typically served by only one psychrometer in several hundred cubic metres of space. Shtienberg and Elad found that over three years, a rain forecasting system did not enable BOTMAN to perform significantly better than a weekly fungicide insurance programme in unheated tomato and cucumber crops.

However, a 4-day weather forecast proved more useful than immediate past records of weekly averages of surface wetness (calculated from dewpoint) of 7 h/d and 9.5 h/d at night temperatures between 9 and 21 °C. By the time the requisite data have been collected and analysed, infection has already begun, and is an irreversible action even with the use of fungicides, which act mostly on germinating spores and thus too late to stop infection. Surface wetness is the key factor in all infections, and so its prediction from rates of change in surface and ambient air temperatures combined, by data processor, with simultaneous rates of change in VPD would be more timely in the immediate application of environmental control measures. Powdery mildew epidemics have a somewhat less complicated sequence of events prior to infection than grey mould epidemics but they, too, are ultimately dependent on the deposition of dew. Control of any fungus-incited disease is achieved by breaking any of the pathways in life cycles similar but the denial of water to germinable spores is the most important. Computer models can be used to optimize greenhouse climate for both crop production and pest and disease control. For example, in The Netherlands a climate management programme was developed for optimal production of tomatoes and is linked to a model for biological control of greenhouse whitefly by *Encarsia formosa* Gahan. Further, the model can be extended with a humidity management module which prevents the development of fungal diseases. Integration of pest and disease control primarily by manipulating the environment is a highly complex problem. Clarke, in describing a computer-managed system, considered the holistic production system as a six-hierarchy complex of factors in which any change at one level affected the other five levels. Thus, any change in greenhouse climate, whether engineered or not, effects changes in pesticide efficacy, biological control agents, pests and disease vectors, diseases, and ultimately productivity and profit. There are a number of electronic decision support systems for various facets of greenhouse pest and disease control and production strategy.

Jones described an expert system with grower selection of climate set points based on his experience; Jacobson further developed an expert system with pre-set points for tomato production; and Dayan developed TOMGRO that modelled physiological processes in tomato. Only Martin-Clouaire considered disease escape in their model for tomato. Van Roermund, however, described the apposition of a whitefly control model to a production model, to which can be added a disease-avoidance model. Clarke and Jewett described a holistic Harrow Greenhouse Crop Management System (HGCMS) for both greenhouse tomato and cucumber. In addition to providing blueprints for production in which the grower has his own input, HGCMS provides user-friendly diagnoses for diseases, pests, biological controls and physiological disorders. It accepts climate monitoring. In addition, HGCMS allows the grower to enter economic data, and will analyse it for him. Conflict resolution, as far as can be agreed among experts, is a feature of HGCMS but ultimately the grower can accept or reject the advice of HGCMS. The use and analysis of computer models and controls depends, of course, on a reasonable degree of computer literacy among growers, together with a basic understanding of plant growth and pest and disease biology. Otherwise reliance on expert advisory services is obligatory.

4

Biological Control and IPM in Protected Crops

Chemical control of pests is not easy and is always accompanied by the risk of development of resistance to pesticides. Also, the application of biological control is more difficult than in Western Europe. Most of the greenhouse pests in Japan are cosmopolitan. They are small in size and suck fluids from plants. They belong to several groups of insects or mites, i.e., aphids, whiteflies, thrips, mites, leaf miners and nematodes. Noctuid moth species such as *Spodoptera litura* (Fabricius) (Lepidoptera:Noctuidae) sometimes outbreak in the southwestern region. Another important fact is that several exotic pests have become serious in protected cultivation. After the rapid development of greenhouse cultivation in the 1960s, *Trialeurodes vaporariorum* (Westwood) (Homoptera: Aleyrodidae) and *Thrips palmi* Karny (Thysanoptera: Thripidae) invaded in 1974 and around 1980, respectively. More recently, *Bemisia argentifolii* Bellows and Perring (Homoptera:Aleyrodidae), *Liriomyza trifolii* (Burgess) (Diptera:Agromyzidae) and *Frankliniella occidentalis* (Pergande) (Thysanoptera:Thripidae) were reported as new pests around 1990. These species have established as major pests in protected cultivation in Japan. Another important fact is the high activity of indigenous natural enemies, which sometimes completely suppress the pests in greenhouses after immigrating into greenhouses from surrounding environment.

INTEGRATED PEST MANAGEMENT

For many years, Canadian vegetable producers relied heavily on protective applications of pesticides to reduce the impact of diseases and pests on their crops. In some cases, excessive amounts of chemicals were applied with little or no regard to effects on non-target organisms or contamination of the produce, soil and water.

In the late 1950s, however, scientists, particularly those dealing with control of arthropod pests, began to develop alternative procedures for crop protection, using a minimum of broad-spectrum chemical insecticides in combination with selective chemicals, biological agents, and modified cultural practices. This work stimulated interest in regulation of plant pathogens by biocontrol agents. In addition, by monitoring population levels of pests and pathogens in relation to crop damage, action thresholds were established for use in crop protection programmes as a guide to the timing of control measures. Thus, the present holistic approach to disease and pest control, termed integrated pest management, is evolving and is being widely implemented. Predictive programmes, such as Botcast for onion botrytis leaf blight, Tomcast for tomato early blight, and Blitecast for potato late blight, are resulting in a more precise approach to disease and pest control and are within the spirit of sustainable agriculture. Similarly, action thresholds for application of chemical insecticides to regulate insect pests on such crops as carrot, celery, crucifers, onion, potato, sweet corn and tomato have been or are being developed.

The effective integration of cultural practices and other methods of disease control is illustrated by the development of a system for the control of pea root rot. This system takes into account four pathogenic fungi and their disease cycles, tillage practices, drainage, soil compaction, cultivar susceptibility, green manure crops, and the predisposition of pea to these fungal pathogens by herbicides. A more limited programme for the control of potato late blight has been developed, and an integrated system for the control of cucumber powdery mildew in the greenhouse has been described, along with several ways of manipulating the greenhouse crop and its environment to escape diseases. Thus far, none of these systems considers the simultaneous control of insect and other pests, so the term 'integrated' is only partially valid, but it is encouraging to note that many greenhouse tomato growers have not used chemical pesticides for several years.

These integrated programmes can be improved by the development and use of selective pesticides to minimize the environmental impact of crop protection practices, particularly on indigenous and introduced agents for biological control. Undoubtedly, the most successful selective pesticides developed to date are formulations of the bacterium *Bacillus thuringiensis*

Applications of these formulations will control most foliage-feeding lepidopterous larvae, including those on cruciferous crops, lettuce and celery, and some coleopterous larvae, notably the Colorado potato beetle on potato and tomato. Such applications have no direct effects on non-target, parasitic or predaceous arthropods and do not contribute to pollution of the soil, water or crop.

Furthermore, residues on the crop are not hazardous to humans and other animals. The potential for using viruses, protozoa, fungi and nematodes that kill insects and mites also is promising. Introduced parasitic or predaceous insects and mites are especially successful for control of pests of greenhouse-grown vegetables. Introductions of beneficial insects and mites, in combination with environmental manipulation and hydroponic systems to minimize the occurrence of diseases, have permitted pesticide-free production of greenhouse vegetables, particularly tomato and lettuce, by many growers. Unfortunately, introductions of beneficial insects and mites, and manipulation of those that occur naturally have been exploited very little for control of insect pests of vegetable crops under commercial field conditions.

METHODS TO CONTROL PEST INSECTS AND MITES

Other methods to control pest insects and mites, such as introduction of attractant and repellant materials, application of growth-regulating chemicals, or mass release of pheromones to cause mating disruption, which have been tested or applied to other crop systems, have not appeared promising or have not been assessed for controlling pests on vegetable crops.

Similarly, the release of sterilized males to reduce development of pest species has not proven economically practical for control of insects on vegetables.

In greenhouse vegetable crops, the high degree of environmental control available to growers enables fully integrated programmes to be devised that give good productivity as well as control of insects, mites and diseases. Biological control, using natural and introduced parasites and predators of pathogens and pests, plays a major role in greenhouse crop protection, but it is more difficult to realize in field vegetable crops because of the lack of control over the environment.

Management of viral diseases depends largely on vector control, crop hygiene, resistant cultivars, and regulatory systems of quarantine, inspection and certification.

Monitoring Certain Pests

Monitoring involves some form of surveillance to detect the presence of a disease or pest before economic damage has occurred and in time to economically apply eradicative or protective measures. Two such methods involve visual inspection and trapping. Visual inspection is the simplest method of monitoring and is used for insect and mite pests and for certain diseases, especially in combination with weather data and epidemiological information concerning sporulation, spore dispersal, infection periods and host plant growth stages. Surveillance is done on a regular basis, beginning soon after seeding or planting out, and includes watching for signs of feeding or irregularities of plant growth.

Trapping is necessary for monitoring certain pests; traps may be purchased for certain pests or some may be constructed at home. The type of trap and its location must be suitable for the target species. Sticky strips or glass slides smeared with petrolatum or other adhesive are used for trapping fungal spores, aphids and thrips; coloured pans with water in the bottom are used for aphids; coloured sticky strips or ribbon-like tapes are available commercially for whiteflies in greenhouses; a variety of lights and light traps are used for moths of many types; and a pheromone in a rubber septum is available for such insects as the European corn borer. Determining the density in soil of nematode pests before planting is important for some crops.

Threshold levels have been established for a number of vegetable crops with regard to the rootlesion and root-knot nematodes.

Cultural Practices

After genetic resistance, the most important strategy for managing plant diseases and pests is best thought of as escape or prevention rather than cure. Such measures are aimed at avoiding the conditions that predispose the crop to infection or infestation; for example, selection of healthy seed of high germinability from a reputable source; treating seed if necessary to eradicate pathogens; checking the health of transplants at their source; practicing adequate crop rotation; sowing and transplanting only when soil conditions (temperature, moisture, nutrients, tilth) are adequate; maintaining adequate spacing and row orientation parallel to prevailing winds to permit good crop ventilation; avoiding overhead irrigation when the crop is in a vulnerable stage (*e.g.* at flowering in bean and pea); avoiding working in fields when the foliage is wet with dew or rain; roguing infected plants from seed crops; picking fruit before it is overripe; avoiding mechanical damage and removing field heat as quickly as possible at harvest; controlling weeds that may harbour pathogens and arthropod pests and contribute to humid microclimates in the crop; and controlling insects that may cause infectible wounds and transmit viruses.

Crop hygiene is a major factor in disease escape because it removes many sources of pathogens. Trash piles, contaminated seed trays, pots and stakes, and infected weeds are important sources of wind- and water-borne fungi and bacteria. Personal hygiene and equipment sanitation are also very important. Bacteria and viruses are easily transmitted on hands, clothing, tools and machinery. Tomato mosaic virus, for example, can persist for several months or years on sapdrenched overalls if they are hung dry in a dark closet. In bean fields, bacterial diseases often follow the tracks of workers walking among wet plants.

In nature, populations of insects and mites are regulated primarily by climate-related factors; by predation and parasitism by other arthropods and by other animals, such as birds; by diseases; by preference for food source; by competition with other plant feeders; and by numerous other factors. Because of the significant mortality of insects and mites by predation, parasitism and diseases, the best control practice for many insects and mites is to do nothing that could reduce the impact of these naturally occurring beneficial organisms and other beneficial agents in the environment.

Cultural practices usually influence the impact of pests indirectly. Tillage, crop nutrition, crop rotation, timing of planting and other management practices affect the vigour of the crop plant. A vigourous, healthy plant is more likely to withstand damage from feeding by sap-sucking insects and mites. Attack by a pest sometimes can be circumvented or damage reduced by tillage, by timing of planting and by selection of field conditions. For example, planting cruciferous crops in well-drained soils and delaying planting to reduce exposure of young plants to cold, damp soil conditions will reduce damage by root maggots. The removal of crop residue from the field and other sanitation procedures may reduce infestations of some pests, such as the European corn borer, in the following crop.

INTEGRATED PEST MANAGEMENT STRATEGIES

Integrated Pest Management strategies require detailed studies which can be broken down into three steps:

1. A precise description of pest population dynamics in space and time in order to assess damage thresholds, to determine key points for control (possibly by modelling), and to evaluate control efficiency;
2. A general survey to estimate the variability in the first step between seasons or across a region;
3. The control strategy, including a survey by the grower of population dynamics.

Each of these steps requires particular sampling methods that differ in accuracy: precise measurements for detailed studies, less precise measurements but which can be used on a larger scale for variability evaluation in the second step, and quick and simple methods for final use by the growers.

Pest and disease intensity may be quantified using two different measurements:

1. Estimation of the population size, *e.g.* number of aphids per leaf, number of fungal spores in a cubic meter of air, etc.;
2. Quantification of the injury caused to the host plant, *e.g.* the proportion of leaf tissue infested by larvae, the relative leaf area covered with disease symptoms, etc. The methods should be easy to use, allow rapid estimation, be applicable over a wide range of conditions, and most of all, be accurate and reproducible. This chapter presents some of these methods which may be used for greenhouse crops.

INSECT PESTS

Estimating Insect Numbers in Samples

We will examine different ways of reducing pest assessment time. Methods based on visual abundance indices will be developed in particular, and examples of their application to insect pests will be given. Since many authors have developed methods which may be used with species other than those that they have studied, the references do not always concern greenhouse species. At each step of a study, the spatial distribution of most pests will be very patchy. Sampling plans will thus require numerous data to reach the required level of accuracy. Particular attention should be given to the evaluation of pest densities at each sampling point. This is a bottleneck which will define the "cost" of the sampling.

Methods are often available to reduce the cost of counts, but, except when automatic counting (*i.e.* picture analysis) is possible, these methods lead to a drastic decrease in accuracy. This loss of accuracy is cumulated with the error induced by the sampling itself, to define the final value of the density estimates. Moreover, several methods that are used to accelerate pest counts (*e.g.* field visual observation) underestimate pest densities per area or volume. Such systematic bias, as well as the accuracy of estimates, must be evaluated before using this kind of method.

The most accurate way of obtaining quantitative data on a pest population is to collect the substrate (*e.g.* host plants) and to take these samples to the laboratory where individuals may be isolated and counted under a stereoscopic microscope. As this method is time-consuming, numerous authors have attempted to reduce the time required. The first step is the mechanical extraction of the individuals from the substrate. They can be extracted by washing, by flotation in high-density medium such as saccharose, or by the use of Berlesse-Tullgren funnels. Once the insects are separated from mud, sand or plant fragments, the clean extract can be fractionated into sub-samples. Both steps give results of varying precision, depending on the medium surrounding the insects and the species involved. The time required for counting the insects is reduced two to five times, but is often still too great (*e.g.* up to two hours per leaf, including washing, sub-sampling and counting, for aphids on cucumber).

Collecting insect substrate is not efficient for species such as thrips, which are very mobile. In this case, the extraction has to be made directly in the field, by using sweepnets, direct picking, mouth vacuum devices or vacuum nets like the Dietrick vacuum (D-vac). The numbers are then calculated from the part of the population which can be recovered. This part is often highly variable and the precision of the method is difficult to evaluate. However, successive sampling at the same sites may enable insect density to be estimated with accuracy.

The method of Suber and Le Cren, frequently used to evaluate fish densities in rivers, was adapted to benthic insect counts by Lapchin and Ingouf-Le Thiec, and then by Lapchin to estimate larval and adult coccinellid densities in wheat fields. It has recently been used in cucumber greenhouses by Boll to estimate the number of thrips on leaves after several successive strikes.

Estimating Insect Population Densities

Insect densities may be estimated *in situ,* without collecting samples. The characteristics of the species distribution on the host plants should be taken into account and the most representative leaves or stems should be observed. This method eliminates the laboratory stage of density estimation, but does not significantly reduce counting time. Observation time may sometimes be drastically decreased if rough counting is performed.

This method is useful if the strong systematic under-estimation of the numbers it produces is constant or depends on known parameters. In the previous example of *Coccinella septempunctata* L. in wheat fields, Lapchin developed a "quick visual method", *i.e.* the observer walked within the 25 m^2 sub-plot for 2 min and counted all the adult coccinellids he saw. The numbers obtained by this method correlated well with the density estimated with the Seber and Le Cren method, and allowed the development of a sequential sampling plan for adult coccinellids in wheat fields. However, such a quick method is not automatically appropriate, even for closely related species, as Frazer and Raworth did not find that "walking counts" of adult coccinellids in strawberry fields were reliable. Variance of insect numbers is closely related to the mean. This property implies that variations in numbers may be more easily perceived by observers who follow a geometric rather than arithmetic scale of density. In the field, such orders of magnitude can be easily translated into abundance classes. This was probably the reason why the use of categorical data was soon considered to be a good way of drastically reducing sampling costs.

In the case of aphid populations, precise density estimates of *Myzus persicae* (Sulzer) were related to the proportion of infested leaves in different parts of potato plants. This method was used on other host plants and statistically developed by several authors. As far back as 1954, presence-absence methods were improved by use of a set of abundance classes. Such classes can be purely arbitrary and, for instance, the sampling units distributed into poor, medium or heavy infestation classes. Several authors used more precisely defined classes, according to the number of colonies, their size and their localization. Different kinds of abundance class systems may be developed, according to the insect studied and its environment.

Building a System of Abundance Classes

Three main types of class can be considered. Firstly, there are classes whose limits are defined by the number of individuals that are seen during one sample unit of observation. A logarithmic scale of these limits was first used by Leclant and Remaudière to estimate *M. persicae* densities on peach trees. Another scale which is based on the approximate powers of has successive classes such as: no insect seen, 1 to 3, 4 to 10, 11 to 30, etc. This scale was used by Ferran to evaluate the density of the rose aphid

[*Macrosiphum rosae* (L.)] on rose bushes, by Boll and Lapchin for *Macrosiphum euphorbiae* (Thomas) in tomato greenhouses, and by Lapchin to estimate mummified *Aphis gossypii* Glover on cucumber plants. Secondly, purely qualitative classes, based on size and number of insect patches or on the percentage of contaminated shoots, may be used. Such classes are generally used for large sampling units such as trees.

Finally, there are intermediate systems which are based on the number of sub-units (*e.g.* leaves of a plant) in each class of a set of qualitative classes.

This system was used by Lapchin to evaluate non-mummified *A. gossypii* on cucumber plants and by Boll for *A. gossypii* in open-field melon crops. A visual class system must be both simple and complete. Ease of use depends on the number of classes and therefore there should be a sufficient number to describe accurately the trends of variability in insect density, but not so high as to be difficult to remember. An optimal class number is generally between five and eight. Another condition required for the use of visual classes is that there must be a biological basis for their definition.

To be representative, a qualitative class set must cover the different kinds of patchiness of the species which may be encountered in the field. For example, on cucumber plants isolated colonies of *A. gossypii* a few centimetres in diameter will first develop around winged immigrants (slightly infested leaves). After several days of development, the colonies suddenly spread all over the leaf area (heavily infested leaves). These simple characteristics, which are associated with the size of the leaves that are heavily infested, define the classes of abundance.

Simplicity of the class system determines both the robustness of the results and the time required for field observations. In the example cited above, the observation of one sampling unit takes approximately 30 sec for each cucumber plant. The class system must cover the whole range of insect densities per sampling unit which may be encountered within the observation period and under different conditions. Thus, this range must be evaluated either from previous knowledge or from trials prior to defining the classes.

Calibration of Visual Abundance Classes

The results of visual observations may often be used without any reference to the number of individuals that they represent and, as such, these ranked qualitative data may be analysed using a large set of non-parametric statistic tools. This method has been used, for example, to evaluate the efficiency of biological control of the rose aphid on rose bushes in public gardens.

When more precise data are needed, each visual class must be calibrated by computing the mean and variability of the number of individuals actually present in the sampling units. This step is very timeconsuming because a large set of precise counts must be gathered so as to represent accurately the variability of the situations in which a given class may be chosen.

Classes Using Environmental Descriptive Variables

Calibration of the classes may be viewed as a statistical model having as a response variable the density of aphids, and as a categorical explanatory variable the visual abundance classes. A complex multivariate regression method, "projection pursuit regression", was adapted to this statistical data. Predictions of these models may be further improved by complementary explanatory variables.

This work, including calibration with complementary variables, was performed with four different class systems which were used to evaluate the density of the aphid *A. gossypii* and its parasitoid *Lysiphlebus testaceipes* (Cresson) on cucumber plants in greenhouses.

Two visual methods were used to estimate densities: "the detailed visual method" (DVM) for a leaf, and the "quick visual method" (QVM) for the whole plant. The class sets of DVM and QVM were built according to the apparent numbers of individuals in the observed sampling units. When the QVM was applied to healthy aphids, the four classes were based on the proportion of the area of leaf infested and on the size of the leaves. Precise counts were also made on the same sampling units and used as the response variable, and the data were divided into reference and validation sets. The reference sets were used to develop the regression models, and the validation sets to test their robustness. The choice of complementary explanatory variables was crucial to the development of these regression models. These variables were selected for their influence on the goodness of fit of the models as well as for the time required for collecting.

For example, when using the detailed visual method and when the target population of the model was either healthy or mummified aphids, seven explanatory variables were used:

- The visual class of the leaf;
- The visual class of the non-target population on the leaf;
- The visual classes of the target population on the upper and lower neighbouring leaves on the same plant;
- The vertical rank of the leaf on the plant;
- The number of leaves on the plant;
- The number of leaves infested by the target aphid on the plant.

Such data can be easily gathered during sampling without significant additional cost. QVM sampling of whole plants yielded a mean error of approximately one class per plant (the limits of each class are in the ratio of). The DVM had a mean error of less than one class. The range of the residuals was generally the same for both the reference and the validation data sets, confirming the robustness of these models.

The same method has now been used to calibrate visual class systems for different pest species on vegetable crops in greenhouses (thrips on cucumber plants, aphids on tomato, melon, eggplant and sprout). Each time that a new regression model is tested, particular attention must be given to the development and sampling of the reference data sets, *i.e.* they must include the same combinations of variables which will be used in further field sampling.

TIME SPENT ON EVALUATING INSECT DENSITIES

Reducing the time spent on evaluating insect densities in sampling units has a cost, which is a decrease in the precision of the estimation. However, the time saved allows the observer to increase the number of units taken into account in a sampling plan, and thus to increase the precision of the mean and variance estimates of the density.

The gain in time is greatly increased when visual methods are used (a ratio of 1:10, when compared with precise counts). Mechanical methods, such as washing (ratio of 1:2), are much slower. However, the evaluation of the precision of visual methods requires a time-consuming calibration of the classes, which must be repeated for each study that deals with a different species, a different environment of the insects or a different scale of observation (*i.e.* plant or leaf).

The decision to undertake such work will depend on the chance of building a "good" visual system.

We can summarize the four criteria of this evaluation as follows:

1. Insect density must be highly variable from one sampling unit (often a host plant) to another, and from one sampling date to another (this condition is easily met for numerous phytophagous insects whose densities vary on logarithmic scales from one host plant to another);
2. Most insects must be visible (for instance, such methods cannot be used for certain aphid species mainly located inside rolled leaves);
3. The visual classes must be simple and distinct, *i.e.* their boundaries have to be easily recognizable in the field;
4. These boundaries must be stable in time and space (for example, independent of the host plant growth stage).

The benefit of building such calibrated visual scales depends on which species is being observed and its environment. The scales are particularly useful for most aphid species, as they can reach very high densities and have strongly aggregated distribution patterns. Since the sampling units are not destroyed, crops can be easily monitored and the population dynamics studied separately in different fields.

Such a method permits large-scale surveys. An example is given in figure a set of cucumber greenhouses in Provence (France) was sampled weekly by using visual abundance classes and number modelling. A regular sampling grid was used in every greenhouse and required less than one hour of observation by two people on each sampling occasion.

IPM IN PROTECTED CROPS

Since the number of available natural enemies for biological control is limited and the technical development of the use of other natural enemies is still preliminary, integrated use of natural enemies and other control measures is essential for commercial application. Suitable crops for integrated control using natural enemies can be assumed to be crops attacked by a small number of serious pests, crops for which a low level of pest damage is tolerable, crops requiring insect pollinators. Tomato, strawberry and sweet pepper are three good crops for developing integrated control in Japan. Table shows an integrated control scheme for tomatoes combining use of natural enemies and selective pesticides.

Table. A Scheme of IPM for Tomatoes in Greenhouses in Japan

Period	Control Measures	Pests
Middle of July	Application of	Thrips, Aphids Granules in Soil
Late August	Sulfur Application	Rust Mites
September-October	E. Formosa, D. Isaea	Whiteflies, Leaf Miners
Early October	IGR Application	Noctuid Moths
Middle of January	IGR Application	Thrips
April-May	E. Formosa, D. Isaea	Whiteflies, Leaf Miners

Noctuid moths sometimes cause severe damage in southwestern Japan, but they can be controlled by *Bacillus thuringiensis* Berliner or IGR. Soil application of systemic pesticides and spot treatment with nonselective pesticides are also possible options in integrated control. In Western Europe, integration of the use of natural enemies and selective pesticides is emphasized. However, the greenhouse environment in Japan is less favorable for biological control because of the extreme temperature conditions and incomplete isolation of the greenhouse environment from the surroundings. Integration with other control techniques is needed for more satisfactory biological control. For example, cultural control, including prevention of immigration of pests, removal of host weeds within and outside greenhouses and avoidance of cultivation of host plants near greenhouses, are fundamental for effective control. After infestation, pests increase almost exponentially in the high temperatures from spring to autumn. In such cases, delay of release of natural enemies often results in failure of biological control. Monitoring systems of early infestation by pests in greenhouses are crucial for successful biological control. Use of resistant varieties of crops should also be studied for integrated control. Partial resistance is adequate for integrated control including the use of natural enemies. In conclusion, integration of all possible techniques is needed for the successful and reliable use of natural enemies in Japan.

INTEGRATED PEST MANAGEMENT IN THE FIELD

Some studies have been carried out to develop IPM systems for eggplants in the field. Potential of *O. sauteri* to control *T. palmi* on eggplants was first evaluated by insecticidal check methods in Okayama prefecture. Potted eggplants were introduced into a screen cage, and both *T. palmi* and *O. sauteri* were released. After spraying with fenthion, which controls *O. sauteri* but does not affect *T. palmi*, was sprayed, *T. palmi* increased rapidly. Without application of fenthion, *T. palmi* was kept at a very low level. Similar experiments were carried out in open fields of eggplants. When the eggplants were treated with fenthion, the peak density of *T. palmi* was four times as large as that on the untreated plants.

Further experiments were also conducted using the selective pesticide, pyriproxyfen, which has a control effect on the pupal stage of *T. palmi* but is harmless to *O. sauteri. T. palmi* on eggplants decreased very rapidly after application of this pesticide, due to a combined effect of the pesticide and *O. sauteri* on the *T. palmi* population. On the other hand, when eggplants were treated with a non-selective pesticide, carbaryl, which is very harmful to *O. sauteri* and harmless to *T. palmi, O. sauteri* disappeared immediately, which caused the increase of *T. palmi*. Effectiveness of the IPM program with a combination of naturally occurring *Orius* spp. and selective pesticides was evaluated in eggplant fields in Fukuoka prefecture. In most of the surveyed fields, the number of insecticide applications and the damage on the fruit by *T. palmi* were remarkably reduced in comparison with conventional fields. In these successful IPM fields, indigenous thrips species were abundant when *Orius* spp. colonized the eggplant fields. With regard to vegetation around eggplant fields, *O. sauteri* and *Orius nagaii* Yasunaga (Hemiptera: Anthocoridae) were predominant on white clovers and on paddy rice, respectively. It is likely that surrounding vegetation serves as a main reservoir for *Orius* populations on eggplants. Indigenous natural enemies were well conserved in this IPM system and effectively suppressed other pests such as cotton aphids, spider mites, and broad mites.

CROP EVOLUTION AND HUMAN CIVILIZATION

Agriculture evolved independently in many places on this earth, but the earliest evidence of farming dates 10,000 years ago in present day Iraq For much of the 200,000 or so years prior to agriculture, humans lived as nomadic hunters, gatherers, and scavengers surviving solely on wild plants and animals. Subsequent domestication of these wild plants and animals from their natural habitats launched agriculture, thus radically transforming human societies. This occurred initially in the Fertile Crescent, the Andean region in South America, Mexico, and parts of Asia, but diffused throughout much of the globe. A change from the nomadic lifestyle to farming led us to become community dwellers, eventually spawning the development of languages, literature, science, and technology as people were freed from the continuous daily task of finding food.

Some regions caught on much faster than others, by margins of thousands of years Plants have also evolved or, more accurately, they have been changed rapidly by human intervention Every crop plant grown today is related to a wild species occurring naturally in its centre of origin, and progenitors of many of our crops are still found in the wild. Early humans must have tried eating thousands of feral plant species from a pool of a quarter of a million flowering plants before settling down on less than one thousand such species, which were subsequently tamed and adapted to farming. A little over 100 crop species are now grown intensively around the world, with only a handful of them supplying us with most of what we now eat.

Through a process of gradual selection, our ancestors chose a very tiny section of the wild plant community and transformed it into cultivated crops. Some profound alterations in the plant phenotype occurred during such selection, and these include determinate growth habit; elimination of grain shattering: synchronous ripening; shorter maturity; reduction of bitterness and harmful toxins; reduced seed dispersal, sprouting and dormancy; greater productivity, including bigger seed or fruit size; and even an elimination of seeds, such as in banana. These changes reduced the survivability of crops in the wild, and thus a feature that transcends all of our crops is the reduction of weedy traits from wild plants. Present crops are thus totally dependent upon human care for their survival, and modern crop varieties would persist in the wild "no longer than a Chihuahua would last in the company of wolves"

Most crops that supply our food were thus obtained at the end of the Stone Age, often from a relatively narrow pool of extant wild genetic diversity. Additional diversity arose within such cultivated crops through new mutations and natural hybridization, and through judicious selection and perpetuation by farmers who maintained them as land races. Varied uses and preferences brought forth further diversification such as in corn (popcorn, sweet corn, dent corn, broom corn, and flour corn for tortilla and corn bread) or the derivatives of ancestral cabbage (kale, kohl rabi, brussels sprouts, cabbage, cauliflower, and broccoli). With the advent of transoceanic navigation and the "discovery" of the New World, crops were moved around the world rapidly, often achieving prominence in adopted homes far beyond their natural centers of origin or domestication.

For instance, the United States is the leading producer of corn and soybean in the word, yet these crops are native to Mexico and China, respectively. The world's largest traded commodity, coffee, had a humble origin in Ethiopia, but now much of it is produced in Latin America and Asia. Florida oranges have their roots in India, while sugarcane arose in Papua New Guinea. Food crops that are now so integral to the culture or diet in the Old World, such as the potato in Europe, chili pepper in India, cassava in Africa, and sweet potato in Japan, were introduced from South America. For that matter, every crop in North America other than the blueberry, Jerusalem artichoke, sunflower, and squash are borrowed from elsewhere! A few sources of our food are also recent domesticates.

Chinese gooseberry occurs wildly in China and is not edible. But careful breeding made it palatable, and it was re-christened "Kiwi fruit" in New Zealand after its introduction there early in the 20th century. The modern strawberry with big fruits is a product of the accidental crossing of two wild species from Virginia (United States) and Chile in France in the mid-18th century. Rapeseed, grown in India for centuries, was altered recently through classical breeding to eliminate the toxic erucic acid and smelly glucosinolates to result in canola—Canadian oil. Triticale, a completely new crop, was artificially sythesized a few decades ago by combining the genomes of wheat and rye (two distinct genera that do not interbreed in nature). It is now grown on over three million acres worldwide. Modern bread wheat itself is also a fairly recent crop in the evolutionary time scale, having arisen only about 4,000 years ago through hybridization of tetraploid (pasta or durum) wheat with inedible goat grass.

FROM MESOPOTAMIA TO MENDEL

While humans have always molded the evolution of crop plants, such changes imposed by farmers occurred over several millennia, leading to rich crop diversity—especially in traits related to their planting or consumption. At the same time, global population grew very slowly until the mid-19th century. It took 1,800 years for the global population to climb from an estimated 300 million around the time when Christianity began, to reach its first billion. But it took only 12 years to add the last billion, rising from five billion people in 1987 to six billion two years ago.

Fortunately, parallel scientific developments in agriculture ensured that food production kept pace with the population explosion of the past century Beginning with Mendel's study of peas, knowledge of genetics helped usher in scientific crop development, resulting in high-yielding varieties. Food production increased in every part of the world in the past few decades, including in Africa. Per capita food consumption has also increased steadily everywhere except in parts of sub-Saharan Africa. In the United States and Canada, where such scientific developments and their applications were most intense, one average farmer now produces enough to feed nearly 150 people! In crops subject to intensive scientific attention—corn, wheat, and rice—the productivity levels increased severalfold. For example, U.S., corn growers averaged 26 bushels of corn per acre in 1928 and 134 bushels per acre in 1998

Such a prodigious increase in agricultural production was underpinned by scientific crop improvement methods along with other developments, including the use of irrigation, improved soil fertility management, mechanization, and control of diseases and pests To develop better crop varieties, scientists have used an array of tools. Artificial crossing, or hybridization, helped us assimilate desirable traits from several varieties into elite cultivars. When desired characteristics were unavailable in the cultivated plants, genes were liberally borrowed from wild relatives and introduced into crop plants. When a crop variety refused to mate with the wild species, various tricks were employed to force them to intermingle, such as the use of the carcinogenic chemical colchicine or by rescuing the hybrid embryos with tissue culture methods. Hybrid vigour was exploited in crops such as corn and cotton to boost productivity. When existing genetic variation within the cultivated germplasm was not adequate, breeders created new variants using ionizing irradiation (gamma ray, x-ray, neutron), mutagenic chemicals (ethyl methane sulfate, mustard gas), or through somaclonal variation

Most people who are concerned about modern biotechnology have little or no knowledge of the processes that have been used to transform crops in the past. Nor are they likely aware that crops have been continually altered over time or that, without human care, they would cease to exist. Using a variety of tools over the past few decades, plant breeders have radically transformed our crop plants by altering their architecture (such as the development of dwarf wheat and rice), shortening growing seasons, developing greater resistance to diseases and pests (all crops), and developing bigger seeds and fruits These crops are also more responsive to management and better adapted to diverse ecological conditions. Improved food quality also resulted through fewer toxins (canola), better digestibility (beans), increased nutrition (high-protein corn), better taste, longer shelf life (thus withstanding long transportation and storage), and enhanced freshness in many vegetables and fruits. A 1,000-fold increase in the marble-sized wild *Lycopersicon* resulted in the modern tomato that can now weigh as much as a kilogram. Modern farming has steadily increased the supply of relatively safe, affordable, and abundant food not only in the developed world, but also in most developing countries.

An average American family now spends only 11 per cent of its income on food and yet has access to better food choices with more variety and nutrition than ever before. Without scientific developments in agriculture, we would otherwise be farming on every square inch of arable land to produce the same amount of food! Using gene transfer techniques to develop GM crops thus can be seen as a logical extension of the continuum of devices we have used to amend our crop plants for millennia. When compared to the gross genetic alterations using wide-species hybridization or the use of mutagenic irradiation, direct introduction of one or a few genes into crops results in subtle and less disruptive changes that are relatively specific and predictable. The process is also clearly more expeditious, as the development of new cultivars by classical breeding typically takes from 10 to 15 years. The primary attraction of the gene transfer methods to the plant breeder, however, is the opportunity to tap into a wide gene pool to borrow traits, obviating the constraints of cross-compatible crop species.

ADDRESSING PUBLIC CONCERNS

While direct gene transfer is still a relatively new approach, many concerns arising from its use may be addressed with the "benchmark" of conventionally bred varieties, as we have the accumulated experience and knowledge with the latter for more than a century. While it seems logical to express a concern such as "I don't know what I am eating with GM foods!" it must be remembered that we really never had that information before with classically bred crops. With GM crops, at least we know what new genetic material is being introduced, so we can test for predictable and even many unpredictable effects. Consider, for example, how conventional plant breeders would develop a disease-resistant tomato.

They would introduce chromosome fragments from its wild relative to add a gene for disease resistance. In the process, hundreds of unknown and unwanted genes would also be introduced, with the risk that some of them could encode toxins or allergens, armaments that wild plants deploy to survive. Yet we never routinely tested most conventionally bred varieties for food safety or environmental risk factors, and they were not subject to any regulatory oversight. We have always lived with food risks, but in the last few decades we have become increasingly more adept at asking questions. To address the concern about long-term health consequences of GM foods, it is instructive to recognize that we worried little about such impacts when massive amounts of new proteins (and unfamiliar chemicals) were introduced into our foods from wild species or when unknown changes were created through mutation breeding. When new foods from exotic crops are introduced, we often assimilate them easily into our diets. What's more we rarely, if ever, before asked the same questions that we now pose about GM crops. Many so-called functional foods, health foods, and nutraceuticals have been entering into the mainstream American diet lately, with little or no regulation or testing. We do not question the long-term health implications of these food supplements, even though they involve relatively large changes in our food intake.

In contrast, the GM foods currently on the market have been tested extensively and judged to be substantially equivalent to their conventional counterparts, with just one or two additional proteins present in miniscule amounts (introduced into a background of thousands of proteins). And, those proteins are broken down either during processing or digestion, with little long-term consequence. In food products such as oil, starch, and sugar, such proteins are not even found. A nagging potential problem with a new protein in food is that it could be a potential allergen. As most food allergens are now well studied, we know that they are found in few defined sources (peanut and other grain legumes, shellfish, tree nuts, and a handful of other foods) and share many similar structural features. Moreover, they must be present in huge proportions in our food, and we must be sensitized to them over time for them to cause any adverse effects. Thus, it is highly unlikely for new allergens to be introduced into our food supply from GM plants.

HISTORICAL ABSENCE

There is no such thing as safe food, and there never has been! That is not to suggest that all of our foods are dangerous, only an acknowledgment that trace levels of such contaminants as toxins and carcinogens are present in everything we eat. But a primary rule of toxicology, articulated over 400 years ago by Paracelsus, refers to the importance of dosage: "Every substance is a poison, but it is the dosage that makes it poisonous" While not alarming, our daily food naturally contains thousands of chemicals, and many of them are shown to be carcinogenic or hazardous in lab animal studies with huge doses.

We consume roughly 5,000 to 10,000 natural toxins daily, as plants have evolved to produce an array of chemicals to protect themselves against pests, diseases, and herbivores 7 have been tested and 19 of them found to be rodent carcinogens The fat-soluble neurotoxins solanine and chaconine are present in potatoes and can be detected in the bloodstream of all potato eaters transferring genes through conventional methods, the resistance is often accompanied by an increase in such toxic compounds. Thus, it is not true that we never had problems with conventionally bred varieties. Any crop variety found to pose a real health risk was promptly removed from the market, but those varieties (in contrast to GM crops) were never routinely tested. One pest-resistant celery variety produced rashes in agricultural workers and subsequently was found to contain 6,200 ppb of carcinogenic psoralens compared to 800 ppb in the control celery This celery was removed from cultivation and that was also the case with the potato variety Lenape, which contained very high levels of toxic solanine. We have always learned from trial and error with all innovations.

Similarly, crop improvement practices evolved over time with continued refinement. It is common, though, for human nature to generate an exaggerated fear of new innovations while perceiving older or "natural" products as always more benign. Huber discusses this double standard in the larger context of risk regulation. We have always been lenient towards existing known and greater hazards, even as we create "gatekeepers" to minimize new risks. Thus, we fail to recognize and "exorcise" much larger older risks.

While most food hazards arise from pathogens such as *Escherichia coli* 0:157, *Listeria monocytogenes,* and *Salmonella enterica* along with mycotoxins produced by fungi (and thus a function of food storage and handling), certain foods containing toxic compounds are known to produce adverse health consequences over time. Cassava, eaten by a large population in Africa, contains cyanogenic glucosides, which cause limb paralysis if consumed before extensive processing. Solanin in tomato and potato is known to cause spina bifida. Vetch pea, a common legume known for its hardiness—and thus popular in India among poor farmers—contains highly dangerous neurotoxins that cause untold misery. Phytohemagglutinin, found in undercooked kidney beans, is toxic. And peach seeds are extremely rich in cyanogenic glucosides. None of these were subject to any mandatory testing before they were introduced into the food chain, nor are they subject to any regulation now. But if the current regulatory standards imposed on GM crops were to be invoked for traditional crops, most of them would fail to meet their requirements.

Humans have built-in natural defences that protect us against normal exposure to toxins. But, according to Ames and Gold we have not evolved to achieve "toxic harmony" with everything we eat, because natural selection occurs much too slowly and because much of what is in our diet today was not eaten at all when we were hunter-gatherers. A balanced mixture of foods normally provides adequate nutrition. However, none of the crops grown today were selected with our nutritional requirements in mind. Instead they were chosen intuitively, by our ancestors, from among the edibles that could be found around them.

Thus, the most important food crop in the developing world—rice—has no provitamin A and little iron in its endosperm. This has led to horrific problems, such as blindness among millions of children due to vitamin A deficiency, and iron-deficiency anemia in nearly a billion women dependent on a rice diet. Biotechnology research, far from causing any new food safety problems, has already demonstrated its potential in enhancing the nutritional quality of our food and is also being employed to reduce harmful toxic compounds that exist in our food.

CROP ROTATION WITH COVER CROPS

One of the biggest challenges of cover cropping is to fit cover crops into your current rotations, or to develop new rotations that take full advantage of their benefits. Some of the systems used successfully by farmers in different regions of the U.S., One might be easily adapted to fit your existing crops, equipment and management. Other examples may point out ways that you can modify your rotation to make the addition of cover crops more profitable and practical.

Whether you add cover crops to your existing rotations or totally revamp your farming system, you should devote as much planning and attention to your cover crops as you do to your cash crops. Failure to do so can lead to failure of the cover crop and cause problems in other parts of your system.

Also remember that there is likely no single cover crop that is right for your farm. Before you start:

- Review Benefits of Cover Crops and Selecting the Best Cover Crops for Your Farm.
- Decide which benefits are most important to you.
- Then consider how these cover crop rotations might be adapted to your particular conditions.
- Talk to your neighbours and the other "experts" in your area, including the contact people listed in Regional Experts.
- Start small on an easily accessible plot that you will see often.

Be an opportunist—and an optimist. If your cropping plans for a field are disrupted by weather or other conditions outside of your control, this may be the ideal window for establishing a cover crop.

Consider using an early-maturing cash crop to allow for timely planting of the cover crop. Cover crops can be used for feed. Consider harvesting or grazing for forage or alternative livestock such as sheep and goats.

COVER CROPS FOR CORN BELT GRAIN AND OILSEED PRODUCTION

In addition to providing winter cover and building soil structure, nitrogen (N) management will probably be a major factor in your cover crop decisions for the corn>soybean rotation. A fall- planted grass or small grain will scavenge leftover N from the previous corn or soybean crop. Legumes are much less efficient at scavenging N, but will add N to the system for the following crop. Legume/grass mixtures are quite good at both.

Corn>Soybean Systems

Keep in mind that corn is a heavy N feeder, soybeans benefit little, if at all, from cover crop N and that you have a shorter time for spring cover crop growth before corn than before soybeans.

Precaution. If you use herbicides, be sure to check labels for plant-back or rotation intervals to ensure that your cover crop isn't adversely affected.

Cover crop features: rye provides winter cover, scavenges N after corn, becomes a long-lasting (6 to 12 week) residue to hold moisture and suppress weeds for your soybeans; hairy vetch provides spring ground cover, abundant N and a moderate-term (4 to 8 week) mulch for a no-till corn crop; field peas are similar to vetch, but residue breaks down faster; red clover is also similar, but produces slightly less N and has less vigourous spring regrowth; berseem clover grows quickly to provide several cuttings for high-N green manure. Here are some options to consider adapting to your system:

Corn>Rye>Soybeans>Hairy Vetch

In Zone 7 and warmer, you can grow a cover crop every year between your corn and full-season beans. Also, you can use wheat or another small grain to replace the cover crop before beans, in a three-crop, two-year rotation (corn>wheat>doublecrop beans). In all cases, another legume or a grass/legume mixture can be used instead of a single species cover crop. Where it is adapted, you can use crimson clover or a crimson/grass mixture instead of vetch.

In cooler areas, plant rye as soon as possible after corn harvest. If you need more time in the fall, try overseeding in rowed beans at drydown "yellow leaf" stage in early fall, or in early summer at the last cultivation of corn. Seeding options include aerial application where the service is economical, using a specialty high-body tractor with narrow tires, or attaching a broadcast seeder, air seeder or seed boxes to a cultivator.

Precaution. Broadcast seeding of cover crops into standing crops is less dependable than other seeding methods. Success will depend on many factors, including adequate rainfall amount and distribution after seeding.

Kill the rye once it is about knee-high, or let it go a bit longer, killing it a couple of weeks before planting beans. Killing the rye with herbicides and no-tilling beans in narrow rows allows more time for cover crop growth, since you don't have to work the ground. If soil moisture is low, consider killing the rye earlier. Follow the beans with hairy vetch or a vetch/small grain mixture.

Legumes must be seeded at least 6 weeks before hard frost to ensure winter survival. Seed by drilling after soybean harvest, or by overseeding before leaf drop. Allow the vetch (or mixture) to grow as long as possible in spring for maximum N fixation.

Harvesting sweet corn, seed corn or silage corn opens a window for timely cover crop planting. Harvest or graze the small grain or legume/small grain mixture in spring if needed for feed.

In Pennsylvania, Ed Quigley seeds rye or spring oats after corn silage harvest. The oats can be cut for silage in fall if planted by early September. Rye can be made into rylage or sprayed before no-till corn the following spring.

Worried about planting your corn a bit late because you're waiting for your cover crop to mature? Research in Maryland, Illinois and elsewhere suggests that no-tilling corn towards the end of the usual window when using a legume cover crop has its rewards. The delay can result in greater yields than earlier planting, due to greater moisture conservation and more N produced by the cover crop, or due to the timing of summer drought. In Pennsylvania, however, delayed planting sometimes reduced corn yields following rye.

Check your state variety trial data for a shorter season corn hybrid that yields nearly as well as slightly longer season corn. The cover crop benefit should overcome many yield differences.

Worried about soil moisture? There's no question that growing cover crops may consume soil moisture needed by the next crop. In humid regions, this is a problem only in an unusually dry spring. Time permitting, allow 2 to 3 weeks after killing the cover crop to alleviate this problem. While spring rainfall may compensate for the moisture demand of most cover crops by normal planting dates, cover crops can quickly dry out a field. Later in the season, killed cover crop residues in conservation tillage systems can conserve moisture and increase yields.

In any system where you are using accumulated soil moisture to grow your cash crop, you need to be extra careful. However, farmers and researchers are finding that water-thrifty cover crops may be able to replace even a fallow year without adversely affecting the cash crop.

Corn>Rye>Soybeans>Small Grain>Hairy Vetch. This rotation is similar to the corn>rye>soybeans rotation, except you add a year of small grains following the beans. In crop rotation research from different areas, many benefits accrue as the rotation becomes longer. This is because weed, disease and insect pest problems generally decrease with an increase in years between repeat plantings of the same crop.

Residue from small grains provides good organic matter for soil building, and in the case of winter grains, the plants help to prevent erosion over winter after soybeans loosen up the soil. The length of the growing season will determine how you fit in cover crops after full-season soybeans in the rotation. Consider using a short- season bean if needed in order to achieve timely planting after soybean harvest. Calculate whether cover crop benefits will compensate for a possible yield loss on the shorter season beans. If there is not enough time to seed a legume after harvest, use a small grain rather than no cover crop at all.

The small grain scavenges leftover N following beans. Legume cover crops reduce fertilizer N needed by corn, a heavy N feeder. If you cannot seed the legume at least six weeks before a hard frost, consider overseeding before leaf drop or at last cultivation.

Precaution. Because hairy vetch is hard seeded, it will volunteer in subsequent small grain crops.

An alternate rotation for the lower mid-South is corn>crimson clover (allowed to go to seed) > soybeans > crimson clover (reseeded) > corn. Allow the crimson clover to go to seed before planting beans. The clover germinates in late summer under the beans. Kill the cover crop before corn the next spring. If possible, choose a different cover crop following the corn this time to avoid potential pest and disease problems with the crimson clover.

Precaution. In selecting a cover crop to interseed, do not jeopardize your cash crop if soil moisture is usually limiting during the rest of the corn season! Banding cover crop seed in row middles by using insecticide boxes or other devices can reduce cover crop competition with the cash crop.

3 Year: Corn>Soybean>Wheat/Red Clover. This well-tested Wisconsin sequence provides N for corn as well as weed suppression and natural control of disease and insect pests. It was more profitable in recent years as the cost of synthetic N increased. Corn benefits from legume-fixed N, and from the improved cation exchange capacity in the soil that comes with increasing organic matter levels.

Growers in the upper Midwest can add a small grain to their corn>bean rotation. The small grain, seeded after soybeans, can be used as a cover crop, or it can be grown to maturity for grain. When growing wheat or oats for grain, frost-seed red clover or sweet-clover in March, harvest the grain, then let the clover grow until it goes dormant in late fall. Follow with corn the next spring. Some secondary tillage can be done in the fall, if conditions allow. One option is to attach sweeps to your chisel plow and run them about 2 inches deep, cutting the clover crowns.

Alternatively, grow the small grain to maturity, harvest, then immediately plant a legume cover crop such as hairy vetch or berseem clover in July or August. Soil moisture is critical for quick germination and good growth before frost. For much of the northern U.S., there is not time to plant a legume after soybean harvest, unless it can be seeded aerially or at the last cultivation. If growing spring grains, seed red clover or sweet-clover directly with the small grain.

An Iowa study compared no-till and conventional tillage corn>soybean>wheat/clover rotation with annual applications of composted swine manure. Berseem clover or red clover was frostseeded into wheat in March. Corn and soybean yields were lower in no-till plots the first year, while wheat yield was not affected by tillage. With yearly application of composted swine manure, however, yield of both corn and soybean were the same for both systems beginning in year two of the 4-year study.

Adding a small grain to the corn>soybean rotation helps control white mold on soybeans, since two years out of beans are needed to reduce pathogen populations. Using a grain/legume mix will scavenge available N from the bean crop, hold soil over winter and begin fixing N for the corn. Clovers or vetch can be harvested for seed, and red or yellow clover can be left for the second year as a green manure crop.

Using a spring seeding of oats and berseem clover has proved effective on Iowa farms that also have livestock. The mix tends to favour oat grain production in dry years and berseem production in wetter years. Either way the mixture provides biomass to increase organic matter and build soil. You can clip the berseem several times before flowering for green manure. Precaution.

Planting hairy vetch with small grains may make it difficult to harvest a clean grain crop. Instead, seed vetch after small grain harvest. Be sure to watch for volunteer vetch in subsequent small grain crops. It is easily controlled with herbicides but will result in significant penalties if found in wheat grain at the elevator.

FACTORS LIMITING USE OF NATURAL ENEMIES

Technically, extreme temperatures in greenhouses and the effect of chemicals are two major factors affecting control capability of *P. persimilis* in commercial greenhouses. Monitoring of the early infestation by spider mites to determine the time of introduction of *P. persimilis* is also crucial for the commercial use of the predatory mite because spider mite outbreaks occur in spring more readily than in Western Europe. Delay of release in greenhouse and low temperature condition in winter are two major factors of failure of use of *E. formosa*. Monitoring of the early infestation of whiteflies is of paramount importance, because one of the reasons for failure of the use of *E. formosa* is delay in its release. Trapping of adult whiteflies by yellow sticky traps and check of occurrence of adults on each plant based on binomial sampling were developed as monitoring methods. Most of the insecticides are harmful to *E. formosa* except for some insect growth regulators (IGRs), while many acaricides and fungicides are less harmful. The most important technical factors for commercializing natural enemies is integration of use of natural enemies with other control measures for controlling other pests on a crop. If the control measures for other pests are incompatible with use of natural enemies, practical use of natural enemies is almost impossible.

Social limiting factors in biological control are:

- Low tolerance of pest damage;
- Supply of natural enemies;
- Poor advisory service for farmers;
- Need of registration.

Low tolerance of pest damage is the main reason for difficulty in biological control in flower production. Lack of the system for supplying enough natural enemies of good quality was the limiting factor in Japan, because private companies had been not interested in biological control in Japan until recently. However, 10 species arthropod natural enemies have been registered and sufficient number of imported or indigenous natural enemies is supplied by private companies now. Advisory service for farmers is very important to obtain good performance of natural enemies for promoting biological control.

OTHER CONTROL MEASURES FOR IPM

Although cultural and physical control measures do not have a curative effect after heavy infestations of greenhouse pests, their preventive effect on the rapid increase of pests is of great importance in integrated control. Their effectiveness is related to regulation of life histories or regulation of dispersal. *T. palmi* is a species of subtropical or tropical origin and is considered to be unable to survive in the field in Japan in winter. Suppression of *T. palmi* in greenhouses before the winter should decrease the density in the next season. *B. argentifolii* is also difficult to survive in winter. Most of the greenhouse pests are polyphagous and can infest many kinds of crops and weeds as host plants.

If suitable host crops are cultivated, or host weeds grow, near greenhouses, the pests can migrate from the greenhouse to surrounding fields and *vice versa*. Host weeds growing in greenhouses are also important for the survival of pests after harvesting. Removal of host weeds inside and near greenhouses and avoidance of cultivation of host crops are fundamental conditions for preventive cultural control.

Helgesen and Tauber pointed out three possible routes of introduction of T. vaporariorum into greenhouses:

1. Artificial introduction of the infested seedlings;
2. Infestation from weeds or other crops in the greenhouse;
3. Infestation from host plants outside but near the greenhouse.

This suggests several possibilities for control by prevention of artificial introduction or dispersal of pests. Cultivation of uninfested seedlings in nurseries, removal of host weeds in greenhouses and the use of nets to cover the openings of greenhouses to prevent immigration of pests from outside. These techniques are also applicable to other pests. Plastic films that absorb ultraviolet rays (UVA) are commonly used as covering materials for plastic greenhouses in Japan. One of its important and interesting effects is the prevention of immigration of pests from outside. It is effective against aphids, whiteflies and thrips. Another important physical control measure is the use of coloured traps. Winged aphids and whitefly adults are attracted to yellow and while thrip adults prefer white or blue. Many types of colored sticky traps have been developed and used for mass trapping or monitoring of these pests. Probably, their usefulness consists in the use in supervised control in chemical or biological control. Yellow sticky traps can be used to determine the time of release of *E. formosa* and to evaluate the effectiveness of control.

RESISTANCE OF BITTER GOURD-LOOFAH GRAFTS

The fusarium wilt of bitter gourd caused by *Fusarium oxysporium* Schlechtend:Fr. emend f. sp. *momordicae* Sun and Huang is hard to control in the field. The technique of cleft grafting to a loofah rootstock was developed and adopted in Taiwan. *F. oxysporium* f. sp. *momordicae* can only attack bitter gourd and summer squash, but not other cucurbits that have the potential to be used as resistant rootstock for bitter gourd. The grafting experiments showed that among the various cucurbit plants tested, only pumpkin, bottle gourd and loofah could be used as rootstock for successful grafting with bitter gourd. Loofah was found not only the best candidate as rootstock for having a highest survival rate in plastic houses and displaying vigorous growth in fields but also showed high resistance to the wilt pathogen even when inoculated by artificial root-pruning method. For three years of field cultivation, the combination of bitter gourd scion (variety of Known-You New No. 3) cleft grafted to loofah rootstock (a local line of L39) was the best one to have high bitter gourd yields which reached 61-108 mt/ha. In contrast, the non-grafted bitter gourd or grafted seedlings by whip root grafting only gave a yield from none to 24 mt/ha. This method has been commonly adopted by the growers in Taiwan.

CONTROL OF SOIL-BORNE DISEASES OF TOMATO

The planting area of tomatoes in Taiwan measures about 4,000 ha with the cultivated season during autumn and winter. In recent years, in order to meet the need of the market, cultivation of tomato within protective facilities during summer time has become more and more prevalent in southern Taiwan. Under high temperature and humid conditions, soil-borne diseases such as bacterial wilt caused by *R. solanacearum*, root-knot nematode caused by *M. incognita* and fusarium wilt caused by *Fusarium oxysporium* Schlechtend: Fr. emend f. sp. *lycopersici* (Sacc.) Snyder and Hansen took place and caused severe losses. To eliminate the greatest limiting factor in the production of tomatoes, Tainan Agricultural Improvement Station and Asian Vegetable Research and Development Center (AVRDC) have cooperated to develop practical control measures since 1998, and have obtained some encouraging results. Currently, TA 6 is the main cultivated variety of tomatoes which is resistant to high temperature and tomato rugose yellowing virus (TRYV), but very susceptible to root-knot nematode. With regard to the widespread and persistent characteristics of soil-borne diseases, the researchers focused on screening varieties of eggplants for compatible resistant rootstocks. They found that grafting of tomatoes on eggplant rootstocks, EG190, EG203 or EG219, could provide resistance to nematodes, and increased the yield by 8 per cent to 43.3 per cent. The resistant eggplant rootstocks were grafted with Known-You 301 cultivar of tomato, which was susceptible to bacterial wilt, and bacterial wilt was significantly reduced as well.

GENETIC ENGINEERING

In Taiwan, several measures have been adopted trying to control plant viral diseases. The measures include the application of integrated cultural practices to escape the transmission vectors, the utilization of artificially induced mild strains for cross protection and the generation of transgenic plants resistant to specified viruses by genetic engineering. Among them, the studies on the control of PRSV by genetic engineering have reached the most remarkable achievement.

CP-Transgenic Papaya

PRSV is a major limiting factor for papaya (*Carica papaya* L.) production in tropical and subtropical areas throughout the world. Control measures that have been used to protect papaya plants from PRSV infection, including selection of planting time to avoid the peak of winged aphids, the use of silver mulch to prevent aphids from visiting seedlings, and conventional cross protection. None of these control methods provided a long period of effective protection against PRSV. The CP gene of a local mosaic type strain isolated from Taiwan, designed PRSV YK, has been sequenced, and transferred into papaya via *Agrobacterium*-mediated transformation by Dr. S. D. Yeh at NCHU, and 45 putative transgenic lines were obtained. When the transgenic lines were challenged with PRSV YK by mechanical inoculation, they showed different levels of resistance ranging from delay of symptom development to complete immunity. After the evaluation of resistance under greenhouse conditions, four highly resistant transgenic papaya lines 16-0-1, 17-0-1, 17-0-5 and 18-2-4 were further evaluated under field conditions for their reactions to infection by PRSV and for fruit production. The data obtained from two rounds of field trials displayed that lines 16-0-1, 17-0-1, and 17-0-5 were highly resistant to PRSV. The total fruit yields in the second trials of each line harvested for six months showed a 3.0-3.2-fold increase, and the commercially valuable fruits a 54.3-56.7- fold increase. These results indicated that the CP-transgenic papaya lines have a great potential for control of PRSV in Taiwan.

Control of PRSV by Genetically-engineered Mild Strains

An attenuated strain, PRSV HA 5-1, that infects papaya without conspicuous symptoms was previously obtained by nitrous-acid induction from the severe parental Hawaii strain HA. It has been used successfully for control of PRSV by cross-protection on a large scale in Hawaii and Taiwan. However, strain-specific protection limits its application in Taiwan and other geographic areas. In order to analyze the genomic region responsible for the pathogenicity of PRSV, infectious *in vitro* transcripts of both HA and HA 5-1 were constructed and various virus hybrids between severe parental virus and its mild mutant were generated by recombination at the cDNA level. The genomic region responsible for the attenuation of HA 5-1 was determined, analyzed and compared with parental severe strain HA.

This genomic segment (nt606-3526) of HA 5-1 was transferred to and replaced the corresponding region of the infectious cDNA clone of Taiwan severe strain YK. The pathogenicity of the generated hybrid virus YKMF was demolished and induced symptoms similar to those caused by to HA 5-1. The crossprotection effectiveness of YKMF was evaluated under greenhouse condition. Over 80 per cent of YKMFprotected papaya plants did not show symptom for three months after inoculation with different severe strains.

The result indicated that the attenuated strain YKMF provides much higher degree of cross-protection and is less strain-specific than HA 5-1 and can potentially be used as a better protective strain than HA 5-1 to protect papaya from PRSV severe infection in Taiwan. In another attempt to improve the cross-protection effectiveness of HA 5-1 against heterologous strains, the CP gene of Taiwan type W-strain TW was amplified by RT-PCR (reverse transcription-polymerase chain reaction) and used to replace the CP gene of

HA 5-1 at the cDNA level. The generated recombinant HA 5-1W also induced mild infections on papaya and cucurbits similar to that induced by HA 5-1. Under greenhouse conditions, results of cross-protection tests indicated that HA 5-1W provides a better degree of protection against PRSV-W than HA 5-1 in *Citrullus lanatus* (Thunb.) Matsum and Nakai, and *Cucumis metuliferus* Naudin. The results indicated that these recombinant mild strains could potentially be used to control the important PRSV-W strains in cucurbitaceous crops.

DECISION TOOLS AND TECHNIQUES FOR IPM

Decision tools are techniques for modelling actual systems and are thus simplifications of actual conditions. They have become widely accepted for several reasons. Decision tools or models are less expensive and disruptive than experimenting with the actual systems and can force the decision-maker to analyse the problem in a logical and systematic manner. Decision tools allow managers to ask "what if" questions and evaluate different scenarios. They can also reduce the time needed to make a decision.

On the other hand, models can be expensive and time consuming to develop. The results may be misused and misunderstood because of the complexity of models and because models may use assumptions that oversimplify actual systems. As stated earlier, many tools are available to assist growers in making IPM decisions in the greenhouse. Practical applications of many decision tools in IPM are reviewed in Norton and Mumford. Although none of the many examples presented are specific to greenhouse IPM, the techniques presented can be applied to many greenhouse IPM problems. The following sections discuss the application of decisionmaking tools to IPM in greenhouses.

DECISION TABLES AND TREES

Decision tables and trees are simple yet powerful tools to assist in the decision-making process. These tools can be used to logically and systematically select among alternatives and the structure provided by these tools can give a valuable framework for further investigations.

Decision Matrix

The decision matrix is used to select among alternatives using the addition of utilities rule. Consider a hypothetical situation where one must select a fungicide from three different alternatives. In this example, four evaluation attributes or criteria are established and weights of importance are assigned to each. These weights reflect the beliefs, concerns and experiences of the decision-maker. Each alternative is evaluated and graded on a 0 to 10 scale on how well it satisfies the criterion. The grade is multiplied by the weight and recorded. The alternative with the greatest sum of weighted values across all criteria (chemical 2) is selected. If the lexicographic rule had been used, then chemical 1 would be selected.

Criteria	Weight	Chemical 1	Chemical 2	Chemical 3
Cost	40	10 / 400	10 / 400	7 / 280
Effectiveness	25	10 / 250	9 / 225	10 / 250
Compatibility with Biologicals	20	5 / 100	9 / 180	8 / 160
Days to Harvest	15	5 / 75	8 / 120	5 / 75
Total	100	825	925	765

Rating Scale, R
Excellent =9-10
Good =7-8
Fair =5-6
Poor =3-4
Unsatisfactory =0-2

Rating 5 / 100 R x W

Fig. Decision Matrix to Select among Fungicides

Pay-off Matrix

A pay-off matrix helps the decision-maker economically evaluate alternatives. Pay-off matrices can be used both for decision-making under risk (where the decision-maker knows the probability of occurrence of the outcomes for each alternative) and decisionmaking under uncertainty (whether probabilities are unknown). A possible pay-off matrix is presented for thrips control on sweet pepper in Ontario greenhouses under uncertainty. The pay-off for each combination of alternative and state of nature (an occurrence or situation over which the decisionmaker has little or no control) is included in the matrix. In this example, the states of nature are low, medium and high levels of thrips attack. The do nothing alternative shows the cost of damage caused by the three levels of thrips attack. The other alternatives include both the cost of the control strategy and its ability to reduce thrips levels and damage. The chemical alternative also includes an estimate of yield reduction resulting from crop injury.

Table. Pay-off Matrix for Decision-Making under Uncertainty

Strategy	Level of Thrips Infestation			Medium Outcome in Row	Minimum Outcome in Row	Row Average
	Low	Medium	High			
Do Nothing	-6,000	-18,000	-60,000	-6,000[2]	-60,000	-28,000
Chemical Control	-32,500	-36,200	-49,500	-32,500	-49,500	-39,400
Biological Control	-7,000	-16,000	-32,000	-7,000	-32,000	-18,333

Maximax↑ Maximin↑ Equally Likely↑

With decision-making under uncertainty, the decision-maker can use three different rules for selecting among the strategies. The maximax (optimistic) rule selects the alternative (do nothing) that maximizes the maximum outcome for every alternative. The maximin (pessimistic) rule selects the alternative (biological control) that maximizes the minimum outcome for every alternative.

The equally likely rule finds the alternative (biological control) with the highest average outcome and assumes that each state of nature is equally likely to occur. With a situation where a grower has kept detailed records of thrips levels in the greenhouse, the probabilities of thrips attacks can be calculated. A pay-off matrix can be developed for this situation where the decision is being made under risk.

The expected monetary value (EMV) for alternative i is:

$$\text{EMV}(i) = \sum_{j=1}^{n} \$if * pj$$

where *n is* the total number of outcomes, $ij is the payoff of alternative *i* for outcome *j,* and p_j is the probability of outcome *j.* A risk-neutral grower, who is unconcerned with year to year variations in outcomes, would choose biological control, which has the highest EMV (in our example, the lowest crop loss).

Most growers are more likely to be risk-adverse and choose a strategy that gives acceptable outcomes at high pest levels. In this case, an extremely riskadverse grower would also choose biological control, which has the best outcome under the worst conditions.

Similarly, a risk taker may choose to do nothing, which has the best outcome under the best conditions. Although pay-off matrices help to economically select among alternatives, they do not allow for non-economical criteria to be considered (such as compatibility with *Bombus* spp. pollinators). If a cost can be determined for these criteria, then they should be included in the analysis.

Table. Pay-off Matrix for Decision-Making under Risk

Strategy	Probability of Thrips Level			
	Low (0.2)	Medium (0.5)	High (0.3)	EMV
Do Nothing	–6,000[1]	–18,000	–60,000	–28,200
Chemical Control	–32,500	–36,200	–49,500	–39,450
Biological Control	–7,000	–16,000	–32,000	–19,000[2]

Decision Trees

Many problems consist of sequential decisions over time. When more than one set of decisions are necessary, a decision tree is appropriate. A decision tree is a graphic display of the decision process which indicates decision alternatives, states of nature and their respective probabilities, and pay-offs for each combination of alternative and state of nature. A decision tree is shown for powdery mildew management in figure (state of nature probabilities and pay-offs are not included). Note that as the crop season progresses, the number of options decreases. If the probabilities of powdery mildew occurrence and the pay-offs are added to the tree, the EMV can be calculated for each branch and the best decision determined. Even if probabilities and pay-offs are not known, decision trees are still very useful by laying out all possible options and providing a framework for deciding which options and strategies need further investigation.

Database Systems

Database systems (DBS) consist of a collection of interrelated data and a set of application programmes to access and manipulate the data. The different data items are stored in related files or tables. The application programmes usually provide functions to enter, edit, browse, query and analysis the data. DBS can help solve pest and disease management problems in several ways. First, the development of a DBS can help to better organize and understand the problem. One of the first steps in developing a DBS is to develop a data model. The data model identifies the paths of information flow, specific data items and relationships among data items. DBS can also assist decision-making by storing detailed records of past pest and disease management strategies, along with the outcomes and costs of these strategies. These records can help the decision-maker select among different strategies based upon past results.

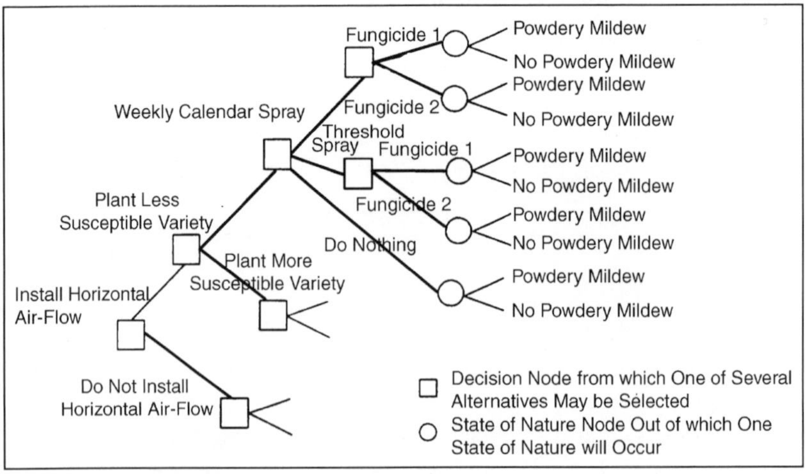

Fig. Decision Tree for Powdery Mildew Management

There are many commercial greenhouse cost accounting and financial management DBS available. These packages usually provide facilities to record costs (including pest control) and sales throughout the cropping season. These DBS can assist in tracking pest management costs in the greenhouse. A DBS for greenhouse pest surveillance, Emerald ICM, is commercially available. Pest survey data is collected with a hand-held computer in the greenhouse and uploaded to a personal computer. The data is used to generate colour maps of pests and their severity over time. Applications of pesticides, fungicides and biologicals are also recorded. Emerald ICM allows the grower to monitor the progress of pest and predator movement and analyse the effectiveness of different control strategies.

Simulation Models

A model is a description of a system. Models may be scaled physical objects, mathematical equations and relations, or graphical representations of actual systems. For purposes of this discussion, a simulation model is a mathematical-logical representation of a system which can be exercised in an experimental fashion on a digital computer. In terms of decision-making, simulation models allow us to examine different alternatives and how these alternatives perform under different conditions.

Simulation models have been used to study various field pests. Pest population growth, fungal passive dispersal, insect active dispersal and predator-prey interactions have been simulated. Pest systems have also been modelled in the greenhouse environment. Nachman simulated the dispersal of two-spotted spider mite (*Tetranychus urticae* Koch) and its predator *Phytoseiulus persimilis* Athias-Henriot in a greenhouse cucumber crop. Spider mite oviposition rate, death rate and emigration rate to other plants are dependent on the health of plants.

The birth, death and emigration rates of the predator are linked to the predation rate. The simulation model was used to study fluctuations in overall population densities within the greenhouse. Biological control of the leafminer species *Liriomyza trifolii* (Burgess) in greenhouse-grown chrysanthemums and *Liriomyza bryoniae* (Kaltenbach) in greenhouse tomato by the parasitoid *Diglyphus isaea* (Walker) have been simulated. Heinz assumed a constant greenhouse temperature of 27 °C and that the population dynamics of the leafminer are independent of the quality of host plant.

Simulation results indicated that successful biological control was unlikely when parasitoid releases are initiated later than 14 d after planting regardless of the release rate. sing a different approach, Boot used ambient temperature and tomato leaf nitrogen content to determine the population dynamics of *L bryoniae.* The timing and growth of leafminer generations were simulated and the results validated with greenhouse experiments, although no practical strategies for parasitoid were given.

These two models could be used to explore different strategies for biological control of leafminers. Disease infection and progression and arthropod pest populations are dependant upon plant nutrition. Crop growth simulation models, such as those developed for greenhouse tomato and cucumber could supply input parameters to pest population models. A feedback loop could potentially predict crop yield reductions due to the pest. Greenhouse microclimate also affects disease development and insect population dynamics.

Microclimate models that simulate the climate within the crop canopy could be combined with disease and pest models to investigate climate control strategies for pest and disease management. Currently, pest management simulation models are being used at the research level to better understand interactions between pests and their control agents.

An example is the simulation model developed by van Roermund to evaluate release strategies for the parasitoid, *Encarsia formosa* Gahan, for control of greenhouse whitefly, *Trialeurodes vaporariorum* (Westwood) on greenhouse crops. As personal computers become more powerful, we envision simulation models being used by growers to evaluate pest management strategies. However, before this happens more research needs to be done to develop models for other pests and diseases. As well, these models must be validated and in a format that non-modellers can understand how to operate the models and interpret the outputs. Model validation can be difficult due to inadequate experimental data, inappropriate assumptions and lack of knowledge regarding some of the physical processes being modelled. Despite these problems, simulation models certainly have a lot of potential for analysing various pest management strategies and understanding interactions between the pest, its biological control agents, the crop and the crop microclimate.

Expert Systems

Expert systems (ES) are computer programmes that emulate the decision-making ability of a human expert. ES contain knowledge in one specific problem area or domain as opposed to knowledge about general problem-solving techniques. ES usually consist of a set of rules that were obtained from an expert to solve a particular problem, and an inference engine that decides which rules to execute. In terms of decision-making, ES can be used as tools for summarizing information and knowledge, selecting among alternative solutions, exploring and evaluating alternative scenarios, assessing risks, diagnosing problems, outlining approaches to problem solving and teaching non-experts the problem-solving approaches of experts. ES technology has been applied to greenhouse pest management. The most common type of application is the diagnosis of crop diseases and pests. Several ES have been developed to identify greenhouse tomato disorders and recommend possible control actions for the identified disorder. With these ES, the user is prompted to enter information about the symptoms displayed by the tomato plant. The ES then uses expert rules to match the observed symptoms with a disorder. Boulard developed an ES to determine the climatic setpoints to control the climate for greenhouse tomato. In addition to information on the outside weather conditions and the current greenhouse climate, the ES also incorporated expert rules on climate and tomato diseases and physiological aspects.

This system attempted to optimize conditions for energy use, crop growth and disease prevention and control. In a separate research effort and using a different approach, Manera developed an ES with similar objectives for greenhouse production and pest management under Mediterranean conditions. Although not specific to greenhouses, other pest management-related ES have been published. Logan developed an expert system to automatically assemble a model describing insect population phenology. The programme offers time savings and compares well with a human expert. Messing describe NERISK, an expert system that assesses the impact of pesticides on beneficial arthropod predators and parasitoids in agricultural systems.

Similar ES could be developed and be useful for greenhouse pest management. Although ES are very good at assisting in the decision-making process, certain issues should be considered before undertaking the development of one. First, it takes considerable time and resources to complete an ES. The time and commitment of an expert(s) is required as well as a knowledge engineer (an individual who extracts, organizes and programmes the knowledge of an expert). The intended user of the ES must be consulted and involved during the development. Ongoing maintenance is required as the knowledge contained in the ES may become outdated. These and other issues relating to ES development are reviewed by Clarke.

CLASSICAL BIOLOGICAL CONTROL OF NATURAL ENEMIES

Classical biological control is the introduction of natural enemies to a new locale where they did not originate or do not occur naturally. This is usually done by government authorities. In many instances the complex of natural enemies associated with an insect pest may be inadequate. This is especially evident when an insect pest is accidentally introduced into a new geographic area without its associated natural enemies. These introduced pests are referred to as exotic pests and comprise about 40 per cent of the insect pests in the United States. Examples of introduced vegetable pests include the European corn borer (*Ostrinia nubilalis*), one of the most destructive insects in North America.

To obtain the needed natural enemies, scientists turned to classical biological control. This is the practice of importing, and releasing for establishment, natural enemies to control an introduced (exotic) pest, although it is also practiced against native insect pests. The first step in the process is to determine the origin of the introduced pest and then collect appropriate natural enemies associated with the pest or closely related species. The natural enemy is then passed through a rigourous quarantine process, to ensure that no unwanted organisms (such as hyperparasitoids) are introduced, then they are mass produced, and released.

Follow-up studies are conducted to determine if the natural enemy becomes successfully established at the site of release, and to assess the long-term benefit of its presence.

There are many examples of successful classical biological control programmes:

One of the earliest successes in the west was in controlling *Icerya purchasi*, the cottony cushion scale, a pest that was devastating the California citrus industry in the late 19th century. A predatory insect *Rodolia cardinalis* (the Vedalia Beetle), and a parasitoid fly were introduced from Australia by Charles Valentine Riley. Within a few years the cottony cushion scale was completely controlled by these introduced natural enemies.

Damage from *Hypera postica* Gyllenhal, the alfalfa weevil, a serious introduced pest of forage, was substantially reduced by the introduction of several natural enemies. 20 years after their introduction the population of weevils in the alfalfa area treated for alfalfa weevil in the Northeastern United States was reduced by 75 per cent. A small wasp, *Trichogramma ostriniae*, was introduced from China to help control the European corn borer making it a recent example of a long history of classical biological control efforts for this major pest. Many classical biological control programmes for insect pests and weeds are under way across the United States and Canada. The population of *Levuana irridescens* (the Levuana moth), a serious coconut pest in Fiji, was brought under control by a classical biological control programme in the 1920s.

Classical biological control is long lasting and inexpensive. Other than the initial costs of collection, importation, and rearing, little expense is incurred. When a natural enemy is successfully established it rarely requires additional input and it continues to kill the pest with no direct help from humans and at no cost. Classical biological control does not always work. It is usually most effective against exotic pests and less so against native insect pests. The reasons for failure are not often known but may include the release of too few individuals, poor adaptation of the natural enemy to environmental conditions at the release location, and lack of synchrony between the life cycle of the natural enemy and host pest.

AUGMENTATION

This third type of biological control involves the supplemental release of natural enemies. Relatively few natural enemies may be released at a critical time of the season (inoculative release) or literally millions may be released (inundative release). Additionally, the cropping system may be modified to favour or augment the natural enemies. This latter practice is frequently referred to as habitat manipulation. An example of inoculative release occurs in production of several crops. Periodic releases of the parasitoid, *Encarsia formosa*, are used to control whitefly, and the predaceous mite, *Phytoseiulus persimilis*, is used for control of the two-spotted spider mite.

Lady beetles, lacewings, or parasitoids such as those from the genus *Trichogramma* are frequently released in large numbers (inundative release). Recommended release rates for Trichogramma in vegetable or field crops range from 5,000 to 200,000 per acre (1 to 50 per square metre) per week depending on level of pest infestation. Similarly, entomopathogenic nematodes are released at rates of millions and even billions per acre for control of certain soil-dwelling insect pests.

The spraying of octopamine analogues (such as 3-FMC) has been suggested as a way to boost the effectiveness of augmentation. Octopamine, regarded as the invertebrate counterpart of dopamine plays a role in activating the insects' flight-or-fight response. The idea behind using octopamine analogues to augment biological control is that natural enemies will be more effective in their eradication of the pest, since the pest will be behaving in an unnatural way because its flight-or-fight mechanism has been activated.

Octopamine analogues are purported to have two desirable characteristics for this type of application:

- They affect insects at very low dosages
- They do not have a physiological effect in humans (or other vertebrates).

Habitat or environmental manipulation is another form of augmentation. This tactic involves altering the cropping system to augment or enhance the effectiveness of a natural enemy. Many adult parasitoids and predators benefit from sources of nectar and the protection provided by refuges such as hedgerows, cover crops, and weedy borders.

Also, the provisioning of natural shelters in the form of wooden caskets, boxes or (turnaround) flowerpots is a form of this. For example, the stimulation of the natural predator *Dermaptera* is done in gardens by hanging up turnaround flowerpots with straw or wood wool.

Mixed plantings and the provision of flowering borders can increase the diversity of habitats and provide shelter and alternative food sources. They are easily incorporated into home gardens and even small-scale commercial plantings, but are more difficult to accommodate in large-scale crop production. There may also be some conflict with pest control for the large producer because of the difficulty of targeting the pest species and the use of refuges by the pest insects as well as natural enemies.

Examples of habitat manipulation include growing flowering plants (pollen and nectar sources) such as Buckwheat near crops to attract and maintain populations of natural enemies. For example, hover fly adults can be attracted to umbelliferous plants in bloom. Biological control experts in California have demonstrated that planting prune trees in grape vineyards provides an improved overwintering habitat or refuge for a key grape pest parasitoid. The prune trees harbour an alternate host for the parasitoid, which could previously overwinter only at great distances from most vineyards. Caution should be used with this tactic because some plants attractive to natural enemies may also be hosts for certain plant diseases, especially plant viruses that could be vectored by insect pests to the crop. Although the tactic appears to hold much promise, only a few examples have been adequately researched and developed.

EXAMPLES OF PREDATORS

Ladybugs, and in particular their larvae which are active between May and July in the northern hemisphere, are voracious predators of aphids such as greenfly and blackfly, and will also consume mites, scale insects and small caterpillars.

Fig. Lacewings are Available from Biocontrol Dealers.

The ladybug is a very familiar beetle with various colored markings, whilst its larvae are initially small and spidery, growing up to 17 mm long. The larvae have a tapering segmented grey/black body with orange/yellow markings and ferocious mouthparts. They can be encouraged by cultivating a patch of nettles in the garden and by leaving hollow stems and some plant debris over winter so that they can hibernate.

Hoverflies resemble slightly darker bees or wasps and they have characteristic hovering, darting flight patterns. There are over 100 species of hoverfly whose larvae principally feed upon greenfly, one larva devouring up to fifty a day, or 1000 in its lifetime.

They also eat fruit tree spider mites and small caterpillars. Adults feed on nectar and pollen, which they require for egg production. Eggs are minute (1 mm), pale yellow white and laid singly near greenfly colonies. Larvae are 8–17 mm long, disguised to resemble bird droppings, they are legless and have no distinct head. Semi-transparent in a range of colours from green, white, brown and black.

Fig. Predatory Polistes Wasp Looking for Bollworms or Other Caterpillars on a Cotton Plant

Hoverflies can be encouraged by growing attractant flowers such as the poached egg plant *(Limnanthes douglasii)*, marigolds or phacelia throughout the growing season.

Dragonflies are important predators of mosquitoes, both in the water, where the dragonfly naiads eat mosquito larvae, and in the air, where adult dragonflies capture and eat adult mosquitoes. Community-wide mosquito control programmes that spray adult mosquitoes also kill dragonflies, thus removing an important biocontrol agent, and can actually increase mosquito populations in the long term. Other useful garden predators include lacewings, pirate bugs, rove and ground beetles, aphid midge, centipedes, spiders, predatory mites, as well as larger fauna such as frogs, toads, lizards, hedgehogs, slow-worms and birds. Cats and rat terriers kill field mice, rats, June bugs, and birds. Dogs chase away many types of pest animals. Dachshunds are bred specifically to fit inside tunnels underground to kill badgers.

More examples:

- *Phytoseiulus persimilis* (against spider mites)
- *Amblyseius californicus* (against spider mites)
- *Amblyseius cucumeris* (against spider mites)
- *Typhlodromips swirskii* (against spider mites, thrips, and white flies)
- *Feltiella acarisuga* (against spider mites)
- *Stethorus punctillum* (against spider mites)
- *Macrolophus caluginosus* (against spider mites)

PARASITOID INSECTS

Most insect parasitoids are wasps or flies. Parasitoids comprise a diverse range of insects that lay their egg on or in the body of an insect host, which is then used as a food for developing larvae. Parasitic wasps take much longer than predators to consume their victims, for if the larvae were to eat too fast they would run out of food before they became adults. Such parasites are very useful in the organic garden, for they are very efficient hunters, always at work searching for pest invaders. As adults they require high energy fuel as they fly from place to place, and feed upon nectar, pollen and sap, thereby pollinating plenty of flowering plants, particularly buckwheat, umbellifers, and composites will encourage their presence.

Four of the most important groups are:

1. *Ichneumonid wasps*: (5–10 mm). Prey mainly on caterpillars of butterflies and moths.
2. *Braconid wasps*: Tiny wasps (up to 5 mm) attack caterpillars and a wide range of other insects including greenfly. A common parasite of the cabbage white caterpillar- seen as clusters of sulphur yellow cocoons bursting from collapsed caterpillar skin.
3. *Chalcid wasps*: Among the smallest of insects (<3 mm). Parasitize eggs/ larvae of greenfly, whitefly, cabbage caterpillars, scale insects and Strawberry Tortrix Moth (*Acleris comariana*).
4. *Tachinid flies:* Parasitize a wide range of insects including caterpillars, adult and larval beetles, true bugs, and others.

Examples of parasitoids: wasp

- *Encarsia formosa* (against white flies)
- *Eretmocerus* spp. (against white flies)
- *Aphidius colemani* (against aphids)

BIOLOGICAL CONTROL WITH MICRO-ORGANISMS

Various microbial insect diseases occur naturally, but may also be used as biological pesticides. When naturally occurring, these outbreaks are density-dependent in that they generally only occur as insect populations become denser.

Bacteria and Biological Control

Bacteria used for biological control infect insects via their digestive tracts, so insects with sucking mouth parts like aphids and scale insects are difficult to control with bacterial biological control. *Bacillus thuringiensis* is the most widely applied species of bacteria used for biological control, with at least four sub-species used to control Lepidopteran (moth, butterfly), Coleopteran (beetle) and Dipteran (true flies) insect pests.

Fungi and Biological Control

Fungi that cause disease in insects are known as entomopathogenic fungi, including at least fourteen species of entomophthoraceous fungi attack aphids. Species in the genus *Trichoderma* are used to manage some soilborne plant pathogens. *Beauveria bassiana* is used to manage different types of pest such whiteflies, thrips, aphids and weevils.

Examples of entomopathogenic fungi:

- *Beauveria bassiana* (against white flies, thrips, aphids and weevils)
- *Paecilomyces fumosoroseus* (against white flies, thrips and aphids)

- *Metarhizium* spp. (against beetles, locusts, Hemiptera, spider mites and other pests)
- *Lecanicillium lecanii* (against white flies, thrips and aphids)
- *Cordyceps* species

COMBINED USE OF PARASITOIDS AND PATHOGENS

In cases of massive and severe infection of invasive pests, techniques of pest control are often used in combination. An example being, that of the emerald ash borer (*Agrilus planipennis* Fairmaire, family Buprestidae), an invasive beetle from China, which has destroyed tens of millions of ash trees in its introduced range in North America. s part of the campaign against the emerald ash borer (EAB), American scientists in conjunction with the Chinese Academy of Forestry searched since 2003 for its natural enemies in the wild leading to the discovery of several parasitoid wasps, namely *Tetrastichus planipennisi,* a gregarious larval endoparasitoid, *Oobius agrili,* a solitary, parthenogenic egg parasitoid, and *Spathius agrili,* a gregarious larval ectoparasitoid. These have been been introduced and released into the United States of America as a possible biological control of the emerald ash borer. Initial results have shown promise with *Tetrastichus planipennisi* and it is now being released along with *Beauveria bassiana,* a fungal pathogen with known insecticidal properties.

5

Detecting Genotoxic Effects of Pesticides in Plants

GENETIC CHANGES IN PESTICIDES

Genetic changes induced by pesticides, their metabolites, and residues are expressed by various endpoints, which include:

- Structural changes in chromosomes and chromatids, called chromosomal aberrations (breaks, deletions, inversions, gaps, translocations, rings) and other disturbances (stickiness, clumping, erosion)
- Disturbances in the mitotic or meiotic division, like spindle inactivation, causing so-called c-mitosis, non-disjunction, and other irregularities in the chromosome distribution during anaphase, resulting in polyploid or aneuploid cells
- Recombinational events like somatic crossing over and sister chromatid exchanges
- Sterility and embryonic lethality
- Mutations in somatic tissues expressed as sectors on leaves and flowers
- Mutation in generative cells expressed either on pollen grains (in haplophase) or in sexual progeny.

To determine these endpoints, a number of plant assays have been developed, based mostly on laboratory application with test compounds, followed by cultivation of plants or their parts (roots) in the laboratory, greenhouse or field. Only a few of them are suitable or have been adapted for testing genetic effects of pesticides *in situ,* under conditions of their utilization in agriculture, horticulture and forestry, or to assess exposures in natural plant communities.

In addition, several plant assay systems frequently used for monitoring genotoxic substance in the air and water as a part of assessing risks to humans have been employed occasionally to test pesticides. As pointed out by Grant (1982a), over 230 plant species have been used in various studies on mutagenesis; however, only a limited number of plant assays are available for routine screening for genotoxicity of pesticides.

CYTOGENETIC ASSAYS

Assays for detection of chromosome aberrations in plants are some of the oldest, simplest, most reliable, and least expensive methods in the field of environmental mutagenesis. Cytogenetic abnormalities can be detected in mitotic and meiotic divisions. Analyses of somatic chromosome aberrations are carried out in cells of growing root tips, stem apexes, or in pollen tubes, whereas studies of meiosis are carried out with pollen mother cells (PMC). Most advantageous for cytogenetic analysis are the species having small numbers of morphologically distinguishable chromosomes that can be viewed microscopically by staining and preparation of squashed slides from soft meristematic tissues. These conditions are met in root tips of the broad bean *Vicia faba* (Kihlman and Anderson, 1984), onion *Allium cepa* (Grant, 1982b), hawks beard *Crepis capillaris, pea* Pisum sativum, in root tips and PMC of barley *Hordeum vulgare,* corn *Zea mays,* and spiderwort *Tradescantia*. This is the reason why these organisms have been used to test for clastogenicity of environmental mutagens, including pesticides. Techniques for staining and slide preparation are so universally known and relatively simple that many other species, including crop and wild plants, may be screened for cytogenetic changes induced by pesticides *in situ* or under laboratory conditions.

Two groups of cytogenetic tests are used in plants:

1. *Laboratory assays:* These are characterized by a short-term (1–24 h) application of test pesticides on seeds or root tips, often at higher doses than used in agriculture, horticulture, or forestry. The evaluation ofchromosomal structural changes is performed during the first cell cycle after pesticide application. Mostly *Vicia faba, Allium cepa,* and barley root tips or seeds are used for this assay.
2. *In situ* assays. They enable application of pesticides under conditions and stage of plant development similar to that in agricultural practice. Experiments of this type can be carried out under field conditions using plants of a single species (mostly barley, corn, *Vicia faba)* or multiple species forming the components of natural plant communities.

Experiments of this type have been carried out by Tomkins and Grant, who analysed 12 different species among 50 detected in experimental plots on roadsides close to agricultural fields. Plants in these plots were sprayed with selected herbicides (picloram, simazine, and diurone) at doses used in agriculture, and their cytogenetic effects were determined in somatic and meiotic cells of flower buds collected at different sampling periods. The researchers observed a significant but differential increase in the frequency of aberrant cells in various species, the highest being in the mid-season of spraying (June and July) and the lowest in the off-season. Two other types of rapid and efficient assays in plants do not detect the cytogenetic damage itself, but rather its consequences: formation of micronuclei, and pollen abortion of sterility and lowered seed setting. The micronucleus test in *Tradescantia is* based on the fact that this organism's chromosomes in early meiotic prophase are extremely prone to breakage; and, subsequently, acentric fragments lead to formation of micronuclei in the tetrade stage of pollen development. Ma (1982) has developed a sensitive assay for scoring micronuclei in flower bud cells after exposition of inflorescences to gaseous or liquid forms of environmental pollutants. The assay is very rapid, because the number of micronuclei can be determined 24–30 h after the action of the tested compound. A special vegetatively propagated clone of *Tradescantia paludosa,*recommended for the assay, can be transferred to the exposed area either as cuttings or as plants in pots. This assay has been applied also for fumigants and several pesticides in the liquid form. Reduced pollen fertility and seed setting are assumed to be consequences of chromosome aberrations induced in the meiotic stage. Indeed, pesticides that induce chromosome aberrations in meiosis also reduce pollen viability (Grant, 1982a).

The pollen sterility assay can be explored for studies in natural plant communities. This kind of research may follow the protocol described by Kurinnyj (1983), based on random sampling of flowers of various species in areas exposed to pesticides and detection of pollen viability by acetocarmine squash technique or by fluorescence techniques. Several test systems in plants are suitable for detection of gene mutations either in many (multiple loci systems) or in one or two specific gene markers (specific locus systems). Mutations are expressed as hereditary changes in the phenotypes observed in haplophase (pollen grains), in somatic tissues of the exposed plants or in the sexual progeny of exposed plants (where mutations of a recessive character are induced in the germ cells of self-pollinated plants).

GENE MUTATIONS SYSTEM

Multiple-loci system tests are very sensitive, because a variety of genetic events in a large number of loci are measured. In the most frequently used systems (*i.e.*, in barley and *Arabidopsis thaliana),* based on determination of chlorophyll and/or embryonic recessive lethals, 700–800 loci are involved (Nilan, 1978). The multilocus system tests in plants are comparable to the *Drosophila* mutation assay, and are considered to be more relevant than specific-locus system tests for prediction of mutagenic response in people. Both in barley and in *Arabidopsis thaliana,* test compounds are usually applied in liquid form on seeds, and the mutations are evaluated after self-pollination and formation of seeds in plants grown in the greenhouse or in the field. Whereas in barley chlorophyll-deficient mutants are recognized at seedling stage in *Arabidopsis* they can be detected on unripe seeds in the siliquae.

The *Arabidopsis* assay is much less time- and space-consuming, and cheaper than the barley assay. In both systems, a large number of mutagenic compounds, including pesticides, have been screened; but seldom *in situ.* A similar protocol to the barley assay is used principally to detect chlorophyll-deficient mutants in wheat, rice, oats, pea, and tomato. Much more experience in the *in situ* monitoring of pesticides is available with the waxy specific locus system in corn. The assay is based on the detection of mutations in the 'waxy' character of pollen grains. In brief, the starch of pollen grains having the dominant wild-type allele *Wx* consists of a mixture of amylose and amylopectine, and it stains bluish-black with iodine. Mutation from dominant to recessive (*wx*) allele leads to the loss of amylose, and the pollen grain stains red. Since over 10^6 pollen grains per treatment group can be scored by a rather simple technique in a short time, this genetic analysis has a statistical power comparable to that obtained in the microbial mutagenesis assays. In the *in situ* experiments, Plewa *et al.* (1984) evaluated the genotoxic activity of pesticides in field plots (2 x 0.5 m or 10 x 3 m) situated randomly within the corn field. Insecticides and herbicides were applied in their field grade formulation at doses similar to that recommended for the control of pests and weed. The waxy system is not specific to corn only, and similar assays have been explored in barley and other species.

Mutations in haploid pollen grains can be explored for the estimation of induced changes in the gene *Adh*$^+$ coding for alcohol dehydrogenase. Grains of corn having *Adh*$^-$ mutations stain yellow, whereas *Adh*$^+$ stains blue.

Another group of specific locus mutation assays in plants, used occasionally to test the genotoxicity of pesticides, is based on specially constructed tester strains, or clones, that are heterozygotic in one or two loci. This enables the manifestation of somatic mutations shortly after the application of a test compound. Somatic mutations are expressed as sectors of differently coloured tissues on leaves (soybean, tobacco, clover, corn) or on flower petals and stamen hairs *(Tradescantia).* The stamen hair assay in *Tradescantia* interspecific hybrids is one of the most sensitive laboratory tests; it has also been used for *in situ* monitoring of gaseous air pollutants, giving the results 2–3 weeks after the exposure of flower buds. The test has been explored also for fumigants (*e.g.*, 1,2-dibromoethane) and pesticides in liquid form (*e.g.*, maleic hydrazide). The tobacco assay, based on the detection of green and white sectors on yellow-green leaves, seems to be very suitable for testing pesticides (*e.g.*, maleic hydrazide) in the soil. This assay, as well as the soybean assay, detects not only reversion or forward mutations, but also somatic recombinations. A tester strain of perennial white clover, with the leaf-colour marker gene, could be a good indicator for long-term monitoring *in situ*, but has never been used for this purpose.

DETECTION OF GENOTOXIC METABOLITES FORMED IN PLANTS FROM PESTICIDES

Formation of mutagenic metabolites from xenobiotics is a well-known phenomenon in animals. Such a mechanism was not considered in plants until it was found to occur in corn sprayed with atrazine and in barley treated with sodium azide. Stable metabolites were isolated, and they caused mutations in microbial tester strains. These observations stimulated research to find suitable and efficient methods to detect mutagenic metabolites in plants formed from environmental compounds including pesticides. Recently, the state of the art in this field has been summarized by Plewa *et al.* (1988).

There are at least three ways to detect the metabolic activation of pesticides into mutagenic products in plants:

1. Exposure of plants or plant-cell cultures to pesticides and testing for mutagenic activity of plant extracts in any mutagenicity assay like *Salmonella,* yeast, or human and rodent cell assay. If extracts from pesticide-treated plants increase the mutagenicity and extracts from unexposed plants are non-mutagenic, the plant genotoxic product can be further purified and characterized. For atrazine's mutagenic metabolite in corn, this can be done by centrifugal fractionation, gel permeation on Sephadex, and HPCL chromatography, whereas for azidoalanin (a mutagenic product of sodium azide in barley), purification and identification are performed by other procedures.
2. Preparation of extracts from untreated plants that retain enzyme activities and their co-incubation with pesticides of interest and with an indicator organism (*e.g.*, bacteria, yeasts, or animals cells). The activation of the pesticide to the mutagenic product mediated by plant enzymes is scored according to the frequency of mutations in the indicator organism. This system has been used successfully by numerous laboratories for pesticide studies to detect the formation of mutagenic metabolites from captan, diquat, maleic hydrazide, triallate, and ziram.
3. Co-incubation of cultured plant cells (instead of plant extract) with the pesticide and with the indicator strain for mutagenicity. This system overcomes many differences associated with the preparation of plant extracts and maintenance of their enzymatic activity.

COMPARATIVE GENOTOXICITY OF THE HERBICIDES ROUNDUP, STOMP AND REGLONE IN PLANT

Pesticides, including herbicides, insecticides and fungicides, are used extensively to improve crop yields and as a result, they accumulate in the environment. More than 2.5 million tons of pesticides are applied every year to agricultural crops worldwide. Pesticides tend to be very reactive compounds that can form covalent bonds with various nucleophilic centres of cellular biomolecules, including DNA. Because of their biological activity, the use of pesticides may cause undesired effects to human health. For instance, the induction of DNA damage can potentially lead to adverse reproductive outcomes, the induction of cancer and many other chronic diseases. Although studies on the biological effects of currently used pesticides have increased in recent years, there are often incomplete, and sometimes contradictory, data on their genotoxicity.

A great variety of tests and test systems based on microbes, plants and animals have been developed in order to assess the genotoxic effects of xenobiotic agents, including pesticides. Arguably, the most reliable genotoxicity evaluation for human health risk is conducted with mammals, whose enzyme systems and more specifically their monooxygenase enzyme complex, are responsible for the biotransformation of xenobiotic chemicals. Although plants have monooxygenase enzyme systems that are to a certain degree similar to the mammal monooxygenase enzyme complex, the plant enzyme complex possesses a number of distinguishing characteristics. Of particular importance are a few reports indicating that unlike animals, some chemicals, including pesticides, are metabolically activated by plant peroxidases and may express different responses compared to those of mammalian cytochromes P-450.

Peroxidases are abundant and widely disseminated in plants and therefore they might play a major role in the plant activation of promutagens. These observations suggest the value of plant test systems for evaluating the genotoxicity of different chemicals used for agricultural purposes. Hence, additional information on the comparative responses of plant and mammalian test systems response to potentially genotoxic pesticides would be of special interest.

In the present investigation, we have evaluated the genotoxicity of three herbicides, Roundup, Stomp and Reglone, in plant (*Crepis capillaris*) and mammalian (mouse) test systems that measure the induction of structural chromosomal aberrations and micronuclei. Chromosomal aberrations qualitatively and quantitatively detect clastogenic activity, while the micronucleus assay detects both clastogenic effects and damage to the mitotic apparatus, some of which might have aneugenic consequences.

STRUCTURE AND FUNCTION OF ECOSYSTEMS

Terrestrial and aquatic ecosystems are generally self-sufficient basic living units of nature that include a complex of species dependent on one another, and interacting with the physical-chemical environment. Pesticides may alter the structure (species richness, density, and biological diversity) and functional activities of an ecosystem. Pesticides may alter the self-sufficient nature of natural ecosystems that include plants (producers), herbivores, parasites and predators, and decomposers.

Some pesticides are capable of destroying some species totally or of significantly reducing the populations of others. When the diversity of the ecosystem is reduced sufficiently, then food chains may be shortened or altered in diverse ways. When food chains are changed, the stability of ecosystems may be reduced, leading a susceptible ecosystem to extinction by a variety of mechanisms such as invasion by other complexes of species.

A critical component of all ecosystems is energy flow. Plants collect energy from the sun for growth, metabolism, and reproduction. Energy fixed by the plants eventually becomes available to herbivores and other species that make up the ecosystem. The more energy the plants collect, the more productive, diverse, and complex is the ecosystem. In aquatic ecosystems, eutrophication—a form of energy cycling—may enhance or limit diversity. If a pesticide influences the growth of the plant populations, then the food/energy supply for the ecosystem is reduced, lowering its productivity and stability.

Another essential component of an ecosystem is the decomposition of organic matter. The basic elements (*e.g.*, C, Ca, H, K, Mg, Mn, N, O, P) are vital to the proper functioning of all life, including ecosystems. Thus, the decomposers are essential to keep the vital nutrients in circulation for use and reuse by an ecosystem.

Some essential nutrients are in the atmosphere. Plants and other organisms may obtain some essential elements (C, H, O, N) from the atmosphere for use throughout an ecosystem. Other nutrients may be obtained directly from soil or water, and are cycled through the biota. Pesticides may be capable of reducing the variability of one or more organisms involved in the recycling process in an ecosystem. If this occurs to a large extent in an ecosystem, it may function at such a reduced rate as to threaten the entire web in the ecosystem.

Methods to quantify ecological effects of pesticides on terrestrial and aquatic ecosystems are generally complex and costly, the magnitude and complexity being dependent on the nature of the pesticides and on the characteristics of the particular terrestrial and aquatic ecosystem in which pesticide is used. Presently, ecosystem parameters are not among those data developed prior to registration of pesticides in most countries. The effects on two ecosystem components seem most urgently to require assessment prior to pesticide registration: these are the effects on the trophic structure (plants, herbivores, parasites, and predators), and on decomposer organisms, because of the unique vulnerability demonstrated by these components for ecosystem integrity and survival. Any new pesticide, whether biological or chemical, should first be assessed for ecosystem effects in the particular agroecosystem of proposed primary use (*e.g.*, an apple orchard or a cotton field). This approach requires the measurement of population dynamics for a representative number of resident species in addition to the target pests and their natural enemies. The number of species to be sampled should be in proportion to the complexity of the agroecosystem of interest. Effects on decomposers can be determined by cellulosic sampling by litterbag methods, for a minimum of three growing seasons.

Effect at Individual and Population Levels

Ecosystems, whether plant or animal, can suffer from accrued damage to individuals. As an illustration, the complexities of such interactions among plants is described.

Plants

Plants are exposed to pesticides, whether as target organisms or as NTOs such as pathogenic fungi. The routes of contact include the uptake from soil and water and deposition via atmospheric drift.

Toxic and mutagenic effects and changes in the metabolism of plants (including formation of metabolites and residues capable of producing adverse effects) can be observed in NTO plants, even with quantitative and qualitative differences resulting from variations in plant sensitivity, and with the numerous metabolic pathways of pesticide degradation. Differential sensitivity of plant species to toxic and genotoxic effects of pesticides has been shown to cause overall changes in species proportion among weeds in croplands and in natural plant communities, because of the reduced abundance of susceptible species and concurrent increases in naturally tolerant species. However, neither the eradication of a susceptible species nor the occurrence of a new plant species has been established in the experimental plots even after long-term (36 years) spraying with formulations such as the herbicide 2,4-D.

The emergence among susceptible plant species of forms resistant to herbicides has become increasingly important. Pesticides can play a dual role as agents favouring the selection of pre-existing resistant mutants and as potential inducers of genetic changes. Genotoxic pesticides are potentially able to increase the pool of mutations in various qualitative and quantitative traits that could cause genetic instabilities of natural plant populations and in crop varieties.

Elements related to possible ecological damage include the formation of stable mutagenic and toxic metabolites and the accumulation of residues in plants that can be harmful to both human and animal populations via exposure through the food chain. This problem needs to be addressed despite the fact that in plants bound residues formed and incorporated into lignin, hemicellulose, and other carbohydrate components of the cell wall are usually less hazardous to the biosphere.

Pesticides have produced changes in both plant metabolism and nutritional patterns which may have further detrimental effects on the ecology. Several mechanisms are known to lead to this consequence.

- Some herbicides, when applied at recommended dosages, have increased the attacks of insect pests and plant pathogens on crops. For example, when corn-growing areas were treated with 2,4-D at the recommended dosage of 1 kg/ha, the numbers of corn leaf aphids increased three-fold; corn borers were 26 per cent more abundant, and were 33 per cent larger than those insects present on untreated corn. Larger corn borers produce one third more eggs, and thus contribute to the build-up of corn borers on corn.

- The insecticides monocrotophos and phosphamidon increased concentrations of nitrogen and phosphorus in rice plants, and these changes were thought to contribute to a resurgence in the numbers of the rice blue leafhopper.
- Herbicides may increase damage due to plant pathogens. For instance, when corn was treated with a recommended dosage of 2,4-D (1 kg/ha), black corn smut grew five-fold larger than on untreated corn. Also, corn that was resistant to southern corn leaf blight lost its resistance to the blight when treated with 2,4-D.
- Most nutrients, especially C, K, N, P, and S, are taken up by plants that, in turn, may be eaten by animals. These nutrients are eventually returned to the soil or atmosphere via decomposition of dead organisms. The amounts and forms of nutrients in soils and plants may be changed by pesticides, thereby altering the dynamics of these animals in the ecosystem.
- Pesticides can alter the chemical make-up of plants. The changes that occur appear to be specific for both the plants and pesticides involved. For example, certain organochlorine insecticides have increased the amounts of some macro and micro-elements (Al, B, Ca, Cu, Fe, K, Mg, Mn, N, P, Sr, and Zn) of corn and beans. DDT, aldrin, endrin, and lindane have stimulated synthesis of the essential amino acids arginine, histidine, leucine, lysine, proline, and tyrosine in corn, but have decreased the content of tryptophan. The herbicide simazine increased water and nitrate uptake in barley, rye, and oat seedlings, resulting in increased plant weight and total protein content.

Increasing insect pest and plant pathogen attacks on crops by using some herbicides may, in turn, lead to an increase in spraying of additional pesticides, like insecticides and fungicides. Thus, the environmental problem arising from the use of herbicides may involve more than just the herbicide itself.

Methods are available to detect both non-genotoxicity and genotoxicity of pesticides in plants, including their effects on plant reproduction. The toxic effects of pesticides have been measured in plants growing in separated experimental plots using endpoints such as frequency of species and individuals in each species, survival of plants, fresh and dry weight of plants at the harvest, and seed setting. Cytogenetic analyses of mitotic cells in roots and shoots, meiotic cells in pollen mother cells, and postmeiotic cells in pollen represent the most frequently used methods to detect genetic changes caused in natural plant populations and communities or in plants growing in the field. Special plant assay systems are also available to monitor genotoxicity *in situ*. About 150–160 pesticides have been tested on plant genotoxicity assay systems. About 90 per cent of the test agents produced some kinds of changes in the chromosome structure, about 60 per cent caused disturbances in the meiosis and seedset reduction, and about 70 per cent caused gene mutations. Compared with other assay systems (*e.g.*, microbial cells, mammalian cell cultures, insects, and whole animals), plant assays appear to be the most efficient to detect pesticide-related genotoxicity. Seasonal variation in cytogenetic endpoints have been correlated with the application of herbicide mixtures.

Animals

The more obvious effects of pesticides (*i.e.*, reproductive failure caused by DDT and mortality caused by OPs and carbamates) have, so far, been the basis of concern and regulatory action.

The major adverse effects of organochlorine (OC) pesticides have been manifested through effects on reproduction. DDT has caused eggshell thinning in several high trophic level avian species and sufficient impact on reproduction to result in population declines (Risebrough, 1986). Likewise, the effects on fish occurred largely during the reproductive cycle (*i.e.*, at the time that the yolk sac was absorbed).

By contrast, the major effect of OPs and carbamates has been direct mortality. The mechanism of action of these insecticides is the inhibition of AChE activity which causes the disruption of nerve function. Acute inhibition of 80 per cent and chronic inhibition of 50 per cent has been associated with mortality. Other esterases are also inhibited, and the inhibition of brain neurotoxic esterase (NTE) has been related to delayed neuropathy (Johnson, 1975). Lotti and Johnson (1978) compared the toxicity of inhibition of AChE and NTE for a range of OPs. The ratio of inhibition of the two enzymes varies over several orders of magnitude, but the degree of inhibition of AChE correlates well with the LD_{50} and that of NTE with delayed neurotoxic damage. For those pesticides causing neurotoxic effects, inhibition of NTE can be one to two orders of magnitude more sensitive than that of AChE.

The relationship between dosage of the pesticide and degree of inhibition of AChE has been used to assess the impact of insecticide spray programme on forest songbirds. The data show a severe collection bias in favour of birds with inhibition of the enzyme of less than 30 per cent. This manifestation is probably due to behavioural effects on birds having AChE inhibition greater than this value.

Since behaviour is the result of integration of many inputs, it has long been considered as a potentially sensitive indicator of pesticide toxicity. For OPs, behavioural alterations are demonstrated only at AChE inhibition of 40–50 per cent. The conclusion that behavioural effects are not particularly sensitive has been confirmed by a review of the effects of toxic chemicals on birds (Peakall, 1985). While it can be stated that behavioural changes will reduce an organism's ability to survive in the wild, little direct evidence exists to confirm this hypothesis. In some cases, predation experiments have been conducted in the laboratory; there remains a major extrapolation to field conditions. Furthermore, in this experiment a significant increase in predation was not seen until the concentration equivalent to half the LD_{50} was used.

GARDENING PESTS CONTROL WITH PLANTS

There are many things on the market today to use as bug repellent for plants. The problems with these repellents are most of them are made with harsh chemicals that are dangerous not only to the pests but to humans as well. Today, people want an organic or natural solution to gardening problems, and few are as natural or effective as repellant planting.

Repellent plants give protection to nearby plants at up to a distance of three feet. Some plants will repel a host of pests while some work against only one or two. It should be noted that planting repellent plants won't get rid of bugs overnight, but it will lead to a pest free environment in time.

Garlic works against most insects and can be planted anywhere without invading the flavour of plants nearby, whether fruits or vegetables. Garlic chives also work with almost any plants. They work great at protecting trees and rose bushes and also look fabulous while doing it, with their white blossoms topping the plants.

The onion family is not alone in its pest repellent capabilities. Certain herbs and flowers serve as great repellent plants also. Tansy (also known as yellow buttons) has a strong fragrance that keeps most bugs at a distance. Mint makes a great companion plant next to cabbage, and it also repels mice. If you have any trouble with mice in your house, leave a few sprigs of fresh mint on your counters at night (or wherever the mice frequent).

Flowers also not only add beauty to a garden, they, too, are great repellent plants. Marigolds are extremely easy to grow and do wonders at keeping bugs at bay, both above and below the soil. The white geranium lures the hungry Japanese beetle, but after eating the plant, they die.

Below is a list of repellent plants and the bugs they help get rid of:

- *Tansy*: Borer, cucumber beetle, cutworm, Japanese beetle, squash bug
- *Garlic/Onion family*: Borer, aphids, mites, rabbits, rose chafer
- *Pennyroyal*: Ants, aphids
- *Mint*: Ants, cabbage maggot, flea beetle, mice
- *Tomato*: Asparagus beetle, flea beetle, cabbage maggot
- *Rosemary*: Cabbage maggot, cabbage moth, carrot fly, Mexican bean beetle, slugs
- *Nasturtium*: Aphids, cabbage moth, squash bug, pumpkin beetle, white fly

- *Sage*: Cabbage maggot, carrot fly, nematode
- *Southernwood*: Ants, cabbage moth, fruit tree moth
- *Wormwood*: Cabbage moth, carrot fly, slugs
- *Marigold*: Eelworm, Mexican bean beetle, tomato hornworm, white fly
- *Petunia*: Aphids, leafhopper, rose chafer, Mexican bean beetle, nematode
- *White geranium*: Japanese beetle, rose chafer, leafhopper
- *Castor bean*: Gophers, moles
- *Green beans*: Colourado potato beetle
- *Potatoes*: Mexican bean beetle
- *Buckwheat*: Wireworm
- *Catnip*: Cabbage moth, flea beetle
- *Asparagus*: Nematode
- *Borage*: Tomato hornworm

Knowing what bugs certain plants repel can help one plan and plant a garden. For example, if the potato beetle hates green beans and a bean beetle hates potatoes, it only stands to reason that one would want to put the two near each other. This holds true for tomatoes and asparagus also. When selecting plants to use as repellents, it is important to study the plant as a whole. For instance, borage may repel the tomato hornworm but at the same time it will attract bees. So before purchasing plants or seeds to fulfill one specific purpose, know as much about them as possible.

A-Z OF GARDEN PESTS

Ants

Studying ants is a lesson in survival. If ants encounter a substance that kills them or forms a barrier, they set to work and pile up enough of each other, dead or alive, to make a bridge to cross. There are over 14,000 ant species worldwide. Some are useful as biological control in farming, many have painful bites and some eat crops. Controlling ants in the garden is the focus here, so whether you have brown coastal ants, bull ants or tiny black ants, most of them continually scout around searching for food and if you see one, it's probably left a scent trail and in a flash the rest of the team will be on their way.

In your vegetable garden ants 'farm' aphids, even moving aphids to put them on better plants. The ants collect the sweet honeydew that aphids secrete after sucking plant sap. Mealy bugs and scale are other soft-bodied insects that secrete honeydew which ants collect. The ants make tunnels and nests in your soil and undermine roots and really roughshod it over everything in your garden if they get out of hand.

You have to be crafty to totally get rid of ants. Some instant ant ammunition... Borax can be used as a natural insecticide, although must be kept away from children, pets and protect yourself too. Watch your eyes, nose, use gloves and wash hands afterwards. Borax mixed with peanut butter or something sweet, such as honey, means the ants eat it and take it back to their nest to share with other ants, hopefully poisoning all in the nest.

Diatomaceous earth sprinkled on ants' trails kills ants by dehydration when they're back in their nest. Garlic fire spray will kill ants on contact... but there's plenty more on their way! There are lots of organic garden pest control deterrent for ants, but keep applying and changing them because they are determined little blighters indeed. When ants are under attack, some ants will gather a few eggs and move to another location, so often all you achieve is for them to move their nest somewhere else, in which case unless they become a pest again, leave them in peace.

Some more ant strategies... Dab some jam, honey or sugar water (and borax if you wish) on the base of aphid infested plants. That will hold the ants back whilst you set some ladybugs on the aphids. You can spray the aphids and ants, but the ants will have a store of eggs underground and will soon move back to start farming again. Cucumber peels on ant routes will send them away for a while. If you find the ants' nest, make them disappear by sprinkling in or near their entrance one or more of these: black pepper, cayenne pepper, cinnamon or chilli powder.

Likewise salt sends them into a frenzy. Boiling water can then be poured on as many ants as you can reach (as long as it's safe and away from plants). Boiling water can of course be poured down an ant nest to kill the queen - if you are lucky enough to find an easy to reach nest, but usually they are very deep and constructed to stop rain and flooded water going in.

Aphids

Rhubarb leaves are semi-poisonous to us, and a tea brewed from rhubarb leaves poisons smaller critters, such as aphids, mites, white fly, caterpillars etc. Pour boiling water over crushed rhubarb leaves then leave to soak for several days. Strain, add a good squirt of detergent and dilute enough so that it looks like weak tea and spray over pest infested plants. Repeat every 10 days or so. Also try the garlic/chilli sprays above.

Birds

Co-dependency between birds and us means we need them and they need us. Wait, don't squawk... bird pest control can be effective, as long as we allow them to control us a bit too. Most birds are insect eaters and will leave plants alone, especially in the breeding season when they are heavy protein eaters.

Okay, it does happen that some birds rip leaves to threads, demolish berries, scatter garden matter far and wide and gouge great holes in your soil. Certain birds, such as pigeons, parrots and lorikeets, can adopt you. Feed one or two and suddenly the sky's full of them all homing into your place!

They don't attack your plants, but can make a mess and ruin railings, porches, roofs and such like.

Poke your beak into some of these friendly, but often surprisingly effective ideas for bird pest control:

- *Shiny things*: String old CDs around the patches of garden where you have planted seedlings for example. Those birds such as blackbirds, that scatter things far and wide, will be wary of the reflective movement.
  Other shiny stuff, such as festive paper and decorations, bottle tops, old utensils, tins and foil are also effective. Use the foil lids from dip and yogurt containers, and don't throw out any foil you used to bake dinner with.
- *Make a scarecrow*: Only feasible if you have the time, space, and especially helpful children.
- *Snakes*: A rubber toy snake or one of those sand filled ones gives excellent pest control. Birds will shy away instead of landing if you put a fake snake in a visible place.
- *Hunters and predatory birds*: A large cutout or kite of a hunter type of bird, such as a hawk or owl, can keep smaller birds away. Nobody likes bullies, even birds. Sit your fearsome picture in a nearby tree or attach to a pole, fence, house or on wire or twine across garden.
- *Cats and dogs*: Having pets doesn't always guarantee the birds will stay away. Anyway, most people that like pets, like birds, so it's back to that mutual understanding of community, jungle type rule of sharing, squaring off and semi-co-existing.

But... here again, a mock puddy tat, especially if larger than life, painted shiny jet black with piecing green or gold eyes, can do the trick. Lean Leo against the nearby fence or pea trellis and move him around often.

How about joining your local night class in woodworking to make your own bird pest control cat - or two?

- *Big bang*: You don't have to create the universe again, just move a few birds along. Make a sudden loud bang or sharp crack. Do this a few times, especially when pigeons are settling down for a nap and they'll get the message and find another guesthouse.
- *Ultrasonic bird repellers*: Unless you're a DIY enthusiast with an electrical bent, then you'll have to buy these online or from a garden centre. These bird repellers emit regular pitched sounds, such as distress calls, predator cries or general scary noises that humans usually can't hear, but will frighten birds and animals away.
- *Bird spikes*: These are probably the most common type of natural bird repellants and popular with boaties to stop seagulls and cormorants from messing up their... well, boats.

For your garden - along fence tops, edges or railings, you can try homemade varieties, but they don't usually look pretty. Nails sticking up from wood, cut plastic or tin - you or your family might get hurt, or goodness - an unwary burglar climbing over your porch rail.

There are lots of types to choose from in shops and online and they can be made from clear, unobtrusive polycarbonate, which doesn't rust or hurt anyone. The birds just can't land or if they do, they get their feet all in a tither.

- *Bird netting*: You can make netting frames to place over fruit and vegetables that are ripening. Frames are a bit of a nuisance to store, but easy to lift and gobble a strawberry or two... or ten.
  Wire or cloth netting is ok to place over plants, but watch out that the plants haven't grown through it and you don't rip them out of the ground when you shift the netting.
  There is some concern that string netting can tangle birds' feet and even friendly snakes, in which case stick with the wire netting or use fine mesh netting attached to high frames.
- *Timed sprinklers*: These are good. You wouldn't hang around either would you, if you got hosed down whilst in the middle of breakfast?
- *Garlic Fire Spray*: Organic garden pest control. This will deter birds if selectively used on the edge of certain precious areas.
- *Harmless sticky gels*: Commercial gels can be bought to use as bird pest control, or make your own either using petroleum jelly or heavy engine oil.

Spread on the oil to keep birds off garden borders, such as wood, bricks etc, or on top of fences, certain branches, gutterings or railings. Birds hate it so it's rather effective, although eventually gets messy.

Caterpillars

To discourage moths and grubs on corn, fruit, brassicas (includes broccoli, spinach, cabbage, cauliflower, Brussels sprouts, kale, collard greens, pak choi and kohlrabi) simply spray with a molasses blend. Mix one tablespoon of molasses with a litre of hot water. Then add a teaspoon of liquid detergent and put in a spray bottle. Spray the leaves, top and bottom, about once a week.

An alternative to the molasses spray is a blend of 1 part vinegar to 3 parts water. Add a teaspoon of liquid detergent and put in a spray bottle. Spray the leaves, top and bottom to kill off grubs and bugs. Also see Garlic fire spray at the top and Rhubarb spray under Aphids.

Cats

Lots of gardeners have cats and that's nice. You can't expect cats to be nice and do what you want them to do, where you want them to do it. They will do what they want to, paw the soil around then have a snooze on top of your seedlings. Cats like a nice bit of loose soil, so mulch like mad and if you can provide a dirt patch or box somewhere else for use, you may be lucky and train them to use that. If you're done being nice or it's your neighbour's cats being bad-mannered - a fright or two with a hose often works. Interestingly a cat that purrs on your lap at your neighbour's house, will be completely different in your garden. Cats are hunters, and they are wary themselves outdoors.

You won't be able to get close enough to hurt them with water, but you can make them unwelcome. A direct hit amidships from 20ft away is harmless, or keep a water filled soda bottle with a hole in the cap handy or child's water pistol to squirt at them. Save your citrus peel, especially orange, and cut it up or run over a pile with the lawnmower. Scatter it around garden edges and keep topping up. Cats thoroughly dislike citrus.

The Garlic fire spray works around garden edges, but you'll go through an awful lot of it, all the time, everywhere you need it, to keep cats away. Miss a corner or forget to spray where they land from the fence next door, and they're in!

Deer

This is a special for those who are trying to grow vegetables around deer. How to repel deer from the garden?

I don't have deer where I live and, hey, they look innocent enough:

- but I can imagine the damage they can do.
- Use the tips listed below to control deer in the garden.
- How to Control Deer in Your Garden
- By Marilyn Pokorney
- Deer are the most difficult of all pests to deter from a garden. They love many different plants. Flowers, vegetables, trees and shrubs are all on their menu.

The best way to keep deer out of a garden is by using a fence that is at least 6 feet high. If it's a small garden that may be feasible but for large gardens or yards, the cost is prohibitive, or possibly not practical. *So the only other choice is deterrents*: Deer don't like the smell of raw eggs, fish products, kelp, or ammonia. Any spray made from these products can be used. Just mix the product of choice in water and spray the plants to be protected. Deer don't like the smell of soap either. Some gardeners have especially found success with Dial and Zest brands. Blood meal scattered around the garden is another good deterrent. Since deer don't like capsaicin, the ingredient that makes peppers hot, a spray made from chili peppers deters deers.Hang fragrant fabric-softener strips, and small nylon bags filled with human hair on trees around the garden.

Some gardeners have found success by laying chicken wire on the ground about six feet wide around the perimetre of the garden. The deer don't like to walk on it because their hooves get stuck in the loops of the wire. Try planting time-released garlic capsules at the bases of trees or shrubs or in the rows of plants to be protected. There is a wide variety of plants that deer won't eat. A list of what they like and don't like can be obtained from your local county cooperative extension office.

Dogs

This is a hairy one. Apart from providing a shady spot somewhere else instead of between your lettuce rows, a soft bit of lawn or soil to roll around on rather than your seedlings, a bone to chew instead of nibbling your peas, and the most horrible face and voice you can conjure up should the bone look like it might be buried in the potato patch... then give up, or put up a fence.

Earwigs

Eeeewahh... If earwigs are eating your plants, crumple some newspaper in old flower pots and leave them in the garden. The earwigs will hide in the paper once they're done eating. In the morning, shake the paper out where they can have a new home or into a bucket of hot water if earwigs aren't your cup of tea.

Fleas

There are whole websites devoted to killing fleas in the home, so here we'll just deal with fleas in the garden. To deter fleas outside use diatomaceous earth and sprinkle it on your shoes and socks as you step into the garden. This way those fleabags won't hitch a ride back inside with you.

DE consists of vast amounts of microscopic fossilized, broken down diatoms. It looks like flour and is harmless to animals and humans. It kills small insects and bugs by cutting into their surface, thus causing dehydration. It's often used in grain silos so is available in farmers' stores as well as some health outlets. It's ok for you and pets to ingest diatomaceous earth as it kills parasites, but it is a very fine powder so avoid breathing it in. Pennyroyal (sometimes called stinking roger) grown near the door and in the garden gets rid of fleas. There are pennyroyal pet soaps and washes that are sold to protect pets from fleas.

Gophers

Here's a few tricks to try to deter gophers:

- Encourage birds of prey by putting water bird baths in garden and keeping foliage around the garden edge down so the birds can spot any such rodents.
- Lay mesh around your garden edge at least 45cm (18") deep. Some gardeners even lay it on the bottom of a garden bed.
- Have some pets, such as a dog or even a cat, and make sure they

are out and about during dawn and dusk at the same time as gophers get active. Gophers and other burrowing creatures will avoid trespassing where there are family pets in sight.

- Spraying with Garlic fire works but depending on garden size it may be too big a task and too expensive to make enough and keep using it.
- If you already have a few gopher holes, put the hose down them and flood their tunnels.

Mealy Bugs

Treat these soft pale pink, white or gray sucking insects the same way as scale. Indoor plants are susceptible to mealy bugs because of the sheltered environment, so in your garden mealy bugs love glass houses and plants growing near or under weather protection. Mealy bugs multiply rapidly and will soon sap the life of a plant, causing distortion and stunted growth first. They are easy to squish with fingers or dab with a cotton bud dipped in methylated spirits. They have a slight waxy coating so anything that burns this off, such as meths, causes death by dehydration. Smothering with oil is also effective.

Mice

If they get into your home from the garden, smarten up your housekeeping to the point of being fanatical. Don't leave food out, put stored food in containers and empty the crumb tray of your toaster. Deter mice with peppermint or eucalyptus oil - soak cloths or cottonwool and put where mice come in or have been. They tend to go around edges and under appliances, so concentrate your efforts there and if you have to lay traps, bait them with peanut butter, cheese, stiff porridge or bread.

Mildew, Mould and Fungus

Plants that are susceptible to mildew, such as peas, pumpkin, zucchini and cucumber can benefit from a milk spray. Use full-cream milk and spray every few days on the leaves until the mildew is under control.

If the small white patches of mildew are just starting, you should be able to catch it early and stop further spreading by using a diluted milk solution of equal parts water and milk. If the situation is out of control, either remove the affected leaves or the whole plant and destroy it.

- Damping off" is caused by fungi such as Phtophtora and Pythium. Seedlings, like lettuce, suddenly collapse with the rot. Soggy conditions, overcrowding or too much peat in the mix can be the cause. Try letting the soil dry out on the surface in-between waterings, and even sprinkling a layer of sand on the surface.

Sand and a dry surface will also control "Fungus gnats." They are tiny little black midget type flies that crawl and fly around the surface of the soil.

They lay their eggs and the larvae will damage seedling by eating their roots. Chamomile tea is a fungicide. Prepare a cup of chamomile tea and allow it to steep for 10 minutes. Once cool it can be sprayed on the affected leaves, top and bottom.

Cinnamon powder sprinkled around plants and soil also controls fungus. If you can make or buy good quality Compost tea, this is extremely beneficial. Compost tea balances the microbes and bacteria in the soil, thus fighting the disease organisms and providing active healthy organisms and fertilizer to strengthen plants and make them resistant to problems.

Moles

See Voles and Gophers. Moles eat insects and worms; they don't harm plants except by their earthworks.

Opossums

Opossum deterrents are many, with only anecdotal evidence that some work. Reliable tests do give credit to camphor to repel opossums. Mix natural camphor flakes or oil to a paste with something like Vaseline and rub around tree trunks, or put camphor in stocking leg or net bag and hang from a branch. If opossums just love your garden and fruit and veggies, it may be worth your while to fence them out. Use floppy but strong chicken netting just under 90cm (3ft) high supported on lengths of fencing wire that are curved outward at the top so those cheeky possums just get swung around when then try to climb up.

Bury the netting 20cm (8") in the ground. For fruit trees put a smooth, wide metal collar around the trunk to stop them climbing up. Keeping chickens nearby may also deter opossums.

Scale: It can be a shock to peer at a plant and realise that it's covered in black, brown or green scale. More of a shock will be when in a short while your plant starts turning up its toes in a stunted sort of way, so tackle scale as soon as you find it. Scale gathers on leaves and stems, sucking the life out of plants.

A black fungus called sooty mould grows on the sweet honeydew secretions from scale. This does the plants no good and looks awful. Ants also love scale honeydew and will aggressively farm scale to obtain as much as they can.

The odd scale can be scraped off with your nail, blunt knife or toothbrush. Always check a week or so later for any new ones. Garlic fire spray also works, because it kills the crawling nymphs as well as smothering the waxy coated adults which permanently attach themselves to the plant.

Also mix 1/4 teaspoon of oil, 2 tablespoons baking soda and 1 tablespoon mild liquid soap in two gallons of water. Spray or wipe on plants once a week for 3 weeks or until gone.

Slugs

Do a night patrol as for snails. The best time is dusk especially when it's damp or raining. Don't use your hand to pick them up - it's horrendous trying to get that sticky stuff off. Scoop them up with an old spoon, dump them in a bucket and scrape them out somewhere where they won't be a pest.

If there's nowhere else for them but kapow... give them a quick stamp with your shoe, or salt sprinkled on snails and slugs will kill them instantly. Go out in the evening with a flashlight and protect your plants! Sprinkle crushed egg shells around vulnerable plants. Slugs and snails will avoid crossing the sharp shells and it will enrich your soil with calcium.

Children like to help here, they love crushing eggshells. Keep them in a jar until you have enough to use. Another dislike of snails is sand, which they do not like to cross. Put a band of fine sand about 1cm (1/4") high around the garden edge or base of plants. Sprays like Garlic fire don't work with slugs and snails - they are mollusks, not insects

Snails

As for slugs mostly. Also put beer in a shallow pan in the garden to trap snails and slugs overnight. Vinegar in a shallow pan will do the same thing.

Orange or grapefruit halves hollowed and turned upside down placed around the garden will also attract snails and slugs. Go out in the morning to remove or deliver the coup de grace on those hiding under the peel.

Spider Mites

There are unknown millions of mite species, many similar to thrips. In the garden the two-spotted mite is commonly called red spider mite because in the cold weather it changes from pale green with two red spots to orange/ red, looking like a miniature red spider.

Like thrips, spider mites suck the chlorophyll from the leaves of plants. Spider mites leave white translucent spots of damage and no dark droppings. In heavier infestations a silky web is obvious, which is secreted by the mites to protect themselves from enemies and poisons. These webs can cover a whole plant.

Natural predatory mites are the best control, so having compost and mulch to encourage beneficial mites will keep the right balance. Spiders, parasitic wasps, lady bugs and lacewings also feed on red spider mites.

Use a soap spray to eradicate these critters. Spray plants weekly until mites are gone, then monthly to stop them from returning. Also see other sprays under Caterpillars.

Thrips

There is no singular word for thrips - such as a thrip. And look at one of their names Heliothrips haemorrhoidalis! So these horribly named beasties come by the gangload. Microscopic greenhouse thrips rasp away on the surface cells of leaves and suck out their chlorophyll contents. The hollowed, air-filled cells then take on a silvery, tissue paper look. This later turns brownish and crinkly and there are dark spots of thrips droppings seen.

Some thrips also spread viruses, such as the tomato spotted wilt virus. Not nice. Keep weeds down if troubled with thrips, as thrips overwinter in nearby ground foliage waiting for their favourite plants the next season. Garlic fire spray trips thrips up and there are a number of predatory bugs native to different countries. Lacewing larvae are a good biological control also.

Voles

Unlike Moles, Voles only eat roots and bark rather than insects and worms. Electronic gadgets were popular once for organic garden pest control, but apparently have little success on Voles, so you might have to resort to mouse traps if you're desperate. Bait traps with bread or stiff porridge, then lay the traps on the vole trails, cover each trap with a box with a hole (to stop birds and possibly other animals, and kids) and keep checking and moving and replacing until you win!

Whitefly

Windy conditions keep whitefly away temporarily, and also a strongish hose down will move them on. Detergent or oil coats their wings and stops them flying, so mix up approximately 1-2 teaspoons detergent to 1 litre (15oz) water, or half a cup of cheap salad oil to 1 litre water, with a dash of detergent. Spray all over infected plants, including underside of leaves if possible.

Whitefly are attracted to yellow. Get some sticky stuff, such as vaseline, castor oil, natural gum resin, or vegetable wax. Tie bright yellow plastic bag bits coated with sticky stuff to stakes, or smear a yellow plastic bucket, or paint a nearby board the brightest yellow you can buy. If necessary shake the whitefly off the plants then watch as they land on the yellow traps and get stuck.

See how Simple Organic Garden Pest Control

You're the owner of those veggies doggone it, not those garden pests. So mix up a batch of whatever you need now and may the luck be with you. Even with natural ingredients, don't get sprays in your eyes, or even touch your eyes with your hands after mixing up concoctions. Label sprays well and keep out of reach of children. Always remember your community spirit towards those that co-share your garden and surrounds. Sometimes you may have to learn to live with a little wastage for you and the planet to stay healthy.

AIM OF SEARCHING FOR HOST-PLANT RESISTANCE

The aim of searching for host-plant resistance or tolerance is to develop cultivars that show little or no reduction in their normal yields when they are exposed to pests and diseases. Growers profit from better yields from resistant crops that need much less use of expensive pesticides and consumers benefit from vegetables with smaller amounts of chemical residues. The capacity of plants to adapt to abiotic and biotic factors was known even to growers in ancient times. When they selected those plants that gave the highest yields and lowest levels of pests and diseases they were unknowingly exploiting genetic resistance. Scientific plant selection started around 1900, and in the following thirty years new varieties with more and more genes of resistance were released. However, subsequent experience revealed that genetic resistance has limits and that sometimes it only serves to combat low pest populations or to delay pest infection; sometimes, resistant cultivars stimulate the selection of pest populations able to live and reproduce on previously resistant cultivars. Consequently, host-plant resistance is best exploited in combination with other techniques like crop rotation, control of weeds within the crops and surrounding areas, biological control of animal pests, etc. Host-plant resistance is then one but important link in the chain of Integrated Pest Management.

TERMINOLOGY

A host plant is a species in which or on which another organism lives. An organism that obtains some advantage from a host plant without benefiting the plant is usually termed a parasite. However, because parasite is used in other chapters of this book for the arthropod species used in biological control, we shall employ the term pest from FAO terminology to denote those weeds, animal species and microorganisms that damage crops. The term pathogen applies to specific microorganisms like bacteria, fungi, mycoplasmas and viruses, that parasitize plants. Plant disorders caused by pathogens are diseases. An animal pest is any animal that usually damages crops (nematodes, insects, mites, etc.). Aggressive strains of a pest are those strains that cause severe symptoms of disease in the plant genotypes attacked. A physiological race of a fungus, bacteria or virus with genes that enable it to attack a specific host-plant genotype is a virulent race; conversely, an avirulent race cannot attack this specific host-plant genotype.

Painter defines host-plant resistance as the relative amounts of heritable characteristics of a plant that influence the degree of damage produced by a pest. Hostplant resistance is then:

- Heritable and controlled by one or more genes;
- Measurable because its magnitude can be determined;
- Relative because measurements are comparative with those of a susceptible plant of the same species that is damaged severely by the pest;
- Variable because it may be modified by biotic or abiotic factors.

If the particularly sensitive phases of plant development do not coincide with the optimum conditions for pest development one speaks about escape.

Against the enormous numbers of pests and plant species in the world, host-plant resistance is common and host-plant susceptibility is exceptional. The combinations of the many types of barriers to infection (resistance characteristics) in a plant species and their collective effectiveness give rise to a series of genotypes that range from highly susceptible to highly resistant. When a pest cannot establish a compatible relationship under any condition with a certain plant genotype, then the genotype is said to be immune or absolute resistant to the pest.

Resistance shown by non-host plants is termed non-host resistance, basic resistance, or basic incompatibility. Non-host resistant plants can exhibit resistance to their specific pests. If a plant expresses some resistance to all isolates or races of a pest it has non-race-specific resistance. If it expresses resistance to only one isolate or pest race it has race-specific resistance. A tolerant plant may be colonized by a pest to the same extent as susceptible plants, but there is no reduction in yield quantity and quality. The converse of tolerance is sensitivity.

Tomato yellow leaf curl virus (TYLCV), for example, produces very mild or no symptoms in both *Lycopersicon chilense* Dun. LA-1969 and *Lycopersicon pimpinellifolium* (Jusl.) Mill. LA-1478, but the concentration of virus antigen in the resistant cultivar LA-1969 is less than 1 per cent of that in the susceptible 'Moneymaker' cultivar, while the concentration in the tolerant cultivar LA-1478 is similar to that in 'Moneymaker'. Rapid recovery of the plant after animal-pest attack is also considered as tolerance.

Resistance Mechanisms

Defence mechanisms present in the plant before pest attack are constitutive mechanisms and those induced by the infection process are induced mechanisms.

Plants do induce responses instead of only constitutive and permanently present resistance because of:

- Chemicals produced by the plant as a result of interactions with pests may be toxic not only for the pest but also for the plant itself leading to a lower plant fitness when no pests attack the plant;
- To produce defence chemicals may be costly, so that plants should allocate resources to defence only when and where interaction with pest occurs.

Constitutive and induced mechanisms may be either morphological or chemical. Examples of morphological constitutive defence mechanisms are the waxes of the cuticule that form a hydrophobic surface preventing water retention and pathogen deposition and germination. Thicker cuticles impede or make difficult penetration of insects, mites and pathogens, particularly when the latter penetrate by appresorium pressure. Thick and tough epidermal cell walls make difficult or impossible direct insect and fungal penetration;

lignification or suberization give additional effective protection. The size and distribution of stomata and lenticels are associated with resistance to those insects, bacteria and fungi that make their entries through these structures. Internal barriers to movement through plant tissues like leaf-vein bundle sheaths and sclerenchyma cells may limit the spread of some pathogens and may prevent penetration of the phloem by aphids and whiteflies. Chemical constitutive defence compounds interfere with the growth and reproduction of pests. The germination of some conidia is inhibited by compounds excreted by the plant. There are also internal secretions of inhibitors like phenolic acids in coloured onions and tomatine in tomato.

Plant tissues may contain antifungal agents produced by normal plant metabolism and, because the concentration of these compounds do not increase in response to infection, they are termed phytoanticipins to distinguish them from the phytoalexins, other chemical defence compounds produced only as a response to infection and mat rapidly reach effective antimicrobial levels around the site of infection. Different plant families produce their characteristically different types of phytoalexins. For example, Fabaceae produce isoflavonoids, and Solanaceae, sesquiterpenes. Furthermore, pest damage can also induce an indirect defence, *i.e.* a defence that improves the effectiveness of natural enemies of the pest. Plants respond to damage by herbivorous mites or insects with the production of volatile chemicals that attract enemies of the herbivore, such as predators or parasitoids.

This plant response occurs both locally and systemically. The morphological and chemical induced defence mechanisms of plants to pests are sometimes associated with the hypersensitive response, a process that leads to the rapid necrosis of infected cells. The pathogen can survive for some time in the necrosed cells around the site of original infection, but the rest of the plant remains healthy. The hypersensitive response is induced by specific elicitors of the pest that interact with specific receptors of the plant (elicitor-receptor model) and, in a number of plant species, it is commonly activated by viruses, bacteria, fungi, insects or mites. The elicitor-receptor model is confirmed in the pathosystem tomato *Cf-9–Fulvia fulva.* (Cooke) Cif. (= Cladosporium fulvum Cooke) race 9. However, in the pathosystem tomato–*Pseudomonas syringae* van Hall pv. *tomato* (Okabe) Young, the hypersensitive response is initiated when the serine-threonine kinase encoded by the resistance gene of the plant interacts physically with the avirulence gene of *Pseudomonas*.

When a virus triggers a hypersensitivity response in a resistant plant, the tissues that surround the necrotic patches develop some localized acquired resistance to further attack by the same or other viruses. The acquired resistance can be shown also by leaves not directly infected by the inductor virus (leaves without hypersensitive necrotic patches) and Ross called this phenomenon systemic acquired resistance. Systemic acquired resistance is not common and even, if present, it does not always protect against a second systemic virus. Pathogen-related proteins and salicylic acid appear to be involved in the mechanism of systemic acquired resistance. Changes in plants after damage by pests or stresses can either decrease or increase plant resistance. The increase in resistance is called induced resistance that is usually systemic and increases with the degree of injury to the plant and reflects complex cytological, histological and physiological changes in the plant. For example, animal pest feeding activities produce short-term responses that affect animal pest feeding behaviour, but also long-term responses that can vary from premature leaf abscission to altered morphology, like increased hair density. Induced resistance elicited by pathogens is also termed cross protection and usually occurs when a plant has been inoculated by a mild strain of the infecting pathogen sometime before the attack of an aggressive strain.

Concurrent protection is a special case of virus cross protection in which the protector virus does not replicate to detectable levels (the plant seems to be immune to that virus), however, the protector virus can induce protection against the second virus. In plants, the two major resistance mechanisms against herbivorous insects are antixenosis (interference with insect behaviour) and antibiosis (interference with insect physiology). The usual patterns of insect approach, landing, probing, feeding and egglaying on a susceptible plant can be disturbed by resistance and induce non-preference or non-acceptance. These disturbances modify the behaviour of the insect and so protect a plant in the initial phase of an attack.

Many examples of plant substances with repellent, deterrent or antifeedant properties are known. Several groups of toxic, secondary plant compounds like alkaloids, flavonoids and terpenoids may adversely affect the growth, development, generation-time and fertility of the insects. Some plant morphological characteristics that can interfere with or modify the behaviour of the insect are colour, shape, type of cuticle wax and the hairiness of plant stalks and leaves.

Genetics of Host-Plant Resistance

The fact that in nature host plants and their pests coexist, even though the pests may sometimes severely damage the plants indicates that they have evolved together and have established a dynamic equilibrium of resistance-virulence. Should either pest virulence or host-plant resistance increase without opposition, then the particular plant or the pest will be eliminated. Consequently, genetic studies of host-plant resistance should include studies of pest virulence genetics.

Inheritance of Resistance

In a segregating plant population, variations of expressed resistance to a particular pest may be either continuous or discontinuous depending on the number of resistance genes involved. Continuous variation from susceptible to resistant plants indicates that the resistance is polygenic which means that it is the sum of the small, individual expressions of many genes. Discontinuous variation indicates that the resistance is monogenic or oligogenic (controlled by one or a few genes) that may be either dominant or recessive major genes: individual plants fall into relatively well-defined classes of resistance or susceptibility.

Genes of resistance are frequently clustered in linkage groups or complex loci and sometimes comprise genes involved in the recognition of more than one taxonomically unrelated pest. The first reported genetic study of resistance was published in 1905. Since then, the enormous amount of work in this field shows that resistance in many cases is inherited in a simple way. Dominance is very common, especially in hypersensitive responses, and recessive resistance occurs much less frequently. Inter-allelic gene interaction (epistasis) is only reported in a few cases.

The Gene-for-Gene Concept

In gene-for-gene relationships, the host-plant resistance expression to a particular pest depends on the pest genotype, and the observed virulence of the pest depends on the host genotype. Flor demonstrates that for each gene that governs resistance in the plant there is a specific gene that governs virulence in the pest. This relationship became known as the gene-for-gene concept and was first shown for a number of fungal diseases and later for viruses, bacteria, nematodes, insects and parasitic plants (*e.g. Orobanche*). Today, it is generally accepted that the interaction takes place between the usually dominant, alleles of the resistance and the, usually dominant, alleles of the avirulence.

The gene-for-gene concept might then be reworded as: any resistance gene can act only if a locus in the pest carries a matching gene for avirulence. Table displays the 16 possible combinations when two genes of resistance in a homozygous diploid plant are matched by two genes of avirulence in the haploid pest. Susceptible plants without no genes of resistance, ($r_1r_1r_2r_2$) are attacked by all races of the pest, even those without genes of virulence (A_1A_2). Pests that carry two genes of virulence (a_1a_2) attack all plants independently of their combinations of genes of resistance. The two pest/plant combinations A_1/R_1 or A_2/R_2 trigger the often hypersensitive resistance response (plant and pest are incompatible).

The combination $a_1A_2/R_1R_1r_2r_2$ is compatible because the avirulence gene A_2 is not matched by the corresponding R_2 host allele. The four possible combinations given by, A_1a_2 and a_2A_1 with R_1r_2 and r_1R_2 illustrate the differential interaction that reveals the occurrence of a gene-for-gene relationship. The differential interaction is used to classify pathotypes and to differentiate genes of resistance.

Table. Interaction of Two Plant Genes of Resistance in a Plant with Two Pest Genes of Virulence According to the Gene-for-Gene Concept

		Resistance (R) or Susceptibility (r) Alleles in the Plant			
		R_1R_1 R_2R_2	R_1R_1 r_2r_2	r_1r_1 R_2R_2	r_1r_1 r_2r_2
A Virulence (A)	A_1A_2	–	–	–	+
Or Virulence (a) Genes in the Past	A_1a_2	–	–	+	+
	a_1A_2	–	+	–	+
	a_1a_2	+	+	+	+

Notes:

- = Incompatibility, + = Compatibility (Infection)

The gene-for-gene interaction produces absolute resistance, or absolute susceptibility, of the host plant against a race of the pest. This race-specific response is termed vertical resistance and is very effective, but only against certain genotypes of a particular pest species. If the resistance is effective against all genotypes of the pest species without differential interaction, the resistance would be race-non-specific or horizontal resistance. The gene-for-gene concept presumably also applies to horizontal (usually polygenic) resistance, although this lacks proof until now.

Durability of Resistance

Johnson and Law proposed the term durable to describe long-lasting resistance. Durability does not imply that resistance is effective against all variants of a pest, but that the resistance has merely given effective control for many years in environmental conditions favourable to the pest. Where susceptible cultivars are grown, the pest population comprises a set of races in dynamic equilibrium, but one or two of the races will tend to predominate. If a resistant cultivar is introduced, the predominant races either will not propagate, or their propagation rate will be substantially less than normal. In both cases, if one or some races can propagate effectively in the resistant cultivar, their proportions in the pest population will increase because they no longer have competition from the other races.

A new outbreak of the pest will occur because the resistance will have been effectively "broken". It is difficult to determine whether a pest population is composed of a mixture of races, some present in very small proportions, or whether the pest produces virulent mutants that disappear from the pest population unless there is a compatible resistant host plant in which they can propagate. In theory, when the introduced resistance is complete, the predominant races will disappear and more virulent races will spread. The spread will be faster than when the introduced resistance is only partial because the virulent and dominant races will compete. Before the introduction of the first resistant tomato cultivars, the predominant if not the only tobacco mosaic virus (TMV) race was race 0. When *Tm-1* resistant cultivars were introduced, the pathogen population changed and very soon TMV race 1 progressively predominated.

Tm-2 cultivars resistant to TMV races 0 and 1 were not much better because TMV race 2 quickly spread. *Tm-1* proved to be resistant to race 2 and cultivars with *Tm-1* and *Tm-2* were released. Again, the resistance of these new cultivars was quickly broken down because TMV race 1-2 predominated. These case histories of *Tm-1*, *Tm-2* and *Tm-1-Tm-2* cultivars support the "lack of durability hypothesis" of complete resistance.

However, the subsequent release of cultivars with the *Tm-2*^{2} allele resistant to TMV races 0, 1, 2 and 1-2 effectively controlled TMV for 20 years. Why the *Tm-1* and *Tm-2* resistances were so ephemeral, and that of *Tm-2*^{2} has lasted more than 20 years, we do not know. Other examples of durable resistances governed by major genes are resistance to *Stemphylium* in tomato and to *Cladosporium* in cucumber. Examples of low durability resistances are those to *F.fulva* in tomato and to *Bremia* in lettuce. Resistance to insects tends often to be partial and polygenic.

It appears then unlikely that more virulent populations (biotypes) adapted to partial resistant cultivars might be selected. However, transgenic cultivars that cany the *Bt* gene from *Bacillus thuringiensis* Berliner rely on a monogenic factor that has a very high expression and, as McGaughey reports, several species of insects like *Helicoverpa zea* (Boddie), quickly adapt to tolerate the *Bt*-gene toxin.

The use of partially resistant cultivars reduces the selection pressures on insect populations and this effectively delays the development of virulent biotypes. The type of reproduction of a pest greatly influences the durability of host-plant resistance.

Aphids, for example, exploit their capacity to reproduce parthenogenetically to colonize resistant cultivars and large populations quickly develop from a few individuals able to overcome host-plant resistance. Soilborne pests, on the other hand, spread more slowly than airborne pests and thus virulent biotypes or races of pathogens take long times to colonize the area in which a resistant cultivar is grown.

Some kinds of host-plant resistance are more durable than others. For example, those which involve changes in plant morphology (growth of hairs or trichomes that interfere with insect movements or feeding activities, or water repellent, waxy surfaces and thickened epidermis of leaves that prevent fungal spores from sticking to the leaf or resist the penetration of some fungi, etc.) require complex changes in the pest to successfully adapt to and overcome the defensive strategy of modified plant structure, and complex changes take very long time.

Breeding to Improve Host-Plant Resistance

Resistant plant varieties are produced by breeding programmes that involve:

- Search for sources of resistance;
- Evaluation of the resistance found;
- Selection in segregating generations.

To growers, the pest resistance of a new variety is only one characteristic out of many, and it is not the most important. Therefore, plant breeders have to bear in mind that the agronomic characteristics of a new resistant variety must be as good as, or better than, previous non-resistant varieties when the pest to which it is resistant is not present. If not, no matter how good its resistance to a particular pest is, the variety is most unlikely to be grown on a large scale.

SOURCES OF RESISTANCE

If the resistance to a particular pest is already present in commercial cultivars (either hybrids or open pollinated cultivars) the source of resistance for our breeding programme would be the resistant commercial cultivar most similar to our ideotype. Commercial cultivars have genes for high yield and quality, for resistances to some pests, for adaptation to specific environments like greenhouses, etc., that must be exploited. Should resistance to the target pest not be present in a commercial cultivar, the first step the plant breeder must take is to search the literature for plants described as sources of resistance, obtain seeds of those source plants, and then evaluate the level of their resistance to help decide whether the source plants might serve as the starting point of the breeding programme.

If the desired resistance is not yet described, it can be searched for in accessions from germplasm banks. The usual search sequence is: landraces, wild forms, related species and related genera. Should it be impossible to find a source of high-level resistance in germplasm collections, the breeding material might still be manipulated by mutation, tissue culture and molecular genetic techniques to produce new variability. Artificially induced mutations have produced a small number of commercial cultivars and, except in those resistances that involve recessive characters in vegetatively propagated ornamental crops, the method is not to be recommended. When cell or tissue cultures are grown for extended periods, genetic variation, termed somaclonal variation, usually takes place.

Examples of useful variation from tissue culture are resistance to *Bipolaris oryzae* (Breda de Haan) Shoemaker and resistance to the herbicide glyphosate. However, in spite of these examples, there are serious doubts about using somaclonal variation as a source of variability, mainly because of the unstability of the variation. To increase the variability of a species by genetic manipulation is limited principally because it is difficult to identify and clone genes. As the number of cloned genes increases, more variability will be generated by plant transformation. The expression of viral DNA sequences in transgenic plants may produce virus-resistant plants that introduce new variability into the gene pool of the plant species.

Evaluation of Resistace

Plant populations must be exposed to the pest in such a way that resistant and susceptible plants can be differentiated as quickly and clearly as possible. Field screening has the advantage that the cost per plant tested is low and, more importantly, that the test conditions simulate those under which commercial crops grow. However, field screening has disadvantages, it is dependent on the weather, whether or not the pest will develop is always uncertain, and other pests may interfere with the tests. Screening under controlled conditions like glasshouses or climatized rooms gives standardized environmental conditions, and the amount of pest present and its distribution can be controlled, but the conditions of growth are not representative of those under which commercial crops grow.

The expression of resistance in a host-parasite system is not constant but it depends largely on the composition and amount of the inoculum, on the stage of development of the plant and on the conditions under which the resistance is evaluated. Small amounts of inoculum produce little or no symptoms in susceptible plants and so resistance may be overestimated. For example, in the pepper–*Phytophthora capsici* Leonian system, concentrations of 10 zoospores/cm^3 of some isolates produce no mortality on 'Morron' cultivar, but concentrations of 10^4 produce 100 per cent mortality. Breeders prefer to test plants as early as possible because seedlings need less space and time to develop and, in general, are less resistant than mature plants. The expression of resistance is greatly influenced by environmental variables (like light, temperature, soil fertility) and the distribution pattern of plant genotypes in the field. To measure resistance properly, the values of those environmental variables should all be within the range of values of the conditions under which commercial crops grow. The expression of resistance shows no constant relationship with light parameters. Host-plant resistance to *Manduca sexta* (Johannsen) in the wild tomato *Lycopersicon hirsutum* Humb. and Bonpl. f. *glabratum* Mull. increases when plants grow under long-day-light conditions, but low-intensity light like that of cloudy days tends to reduce the expression of resistance to insects.

Temperatures outside the range of conditions under which commercial crops are grown reduce the expression of resistance in a number of host-pest systems. However, Gómez-Guillamón and Torés report that three lines of melon, when grown at normal temperatures for commercial crops, show resistance to *Sphaerotheca fusca* (Fr.) Blumer. at 26 °C but are susceptible below 21 °C. High doses of nitrogenous fertilizers generally increase the susceptibility while additional applications of potassium and phosphorus fertilizers increase resistance. When resistance of different genotypes is assessed in small plots, resistant genotypes will export small levels of inoculum, but will receive high levels of inoculum from susceptible genotypes that, in turn, export more inoculum than they receive and, consequently, the resistance of resistant genotypes will tend to be underestimated in comparison with that of the same genotypes measured in trials carried out in large plots or in separate plots.

This phenomenon is termed interplot interference and can be mitigated by including control cultivars with different levels of resistance as references. In any case, small differences found in the level of infection in small plots should be most carefully noted. Pests that generally display a vertical dispersion show smaller interplot interference than pests that display a horizontal dispersion. The principal application of *in vitro* resistance screening is to select those cells, calli, or somatic embryos that show resistance to the toxin of a pathogen.

Advantages of this technique are:

- Large numbers of individuals can be processed;
- Haploid cells reveal concealed recessive traits;
- It can exploit somaclonal variation;
- The uniformity of the experimental environmental conditions helps discriminate slight quantitative differences of plant resistances.

Disadvantages are:

- It is limited to tests for pathogens that produce toxins;
- Cells that survive infection may be physiologically adapted and not genetic variants;
- Resistance at cellular level is not necessarily expressed in the whole plant;
- *In-vitro* resistance screening does not detect defence mechanisms that are based on differentiated tissues.

Selection Methods

After a source of resistance has been identified and an appropriate evaluation procedure has been set up, the next step is to integrate the resistance into the set of agronomic characters that a cultivar needs for success on the market. The donor of resistance should be selected taken into account that the closer the genotype of the donor to that of the cultivar to be improved, the shorter will be the process of introduction of resistance. Complete resistance is frequently easier to manage than partial resistances. Complete resistance is essential for pests damaging the end-product of the crop because the greenhouse-grown produce must be of prime quality without cosmetic damage like spots or scars that reduce consumer acceptability.

Most cultivars of the greenhouse-grown species are hybrids. To produce a resistant hybrid the resistance has to be introduced into one of the parents (dominant resistance), or in both parents (recessive resistance). The appropriate selection procedure for monogenic resistances is backcross and for polygenic resistances is recurrent selection. Marker-assisted selection recovers genes linked to markers. The markers are more easily scored than the genes of resistance. To ensure that only a minor fraction of the individuals selected are recombinants, the linkage between the marker and the target gene in coupling phase should be <5 cM. A repulsion-phase marker linked at <10 cM provides higher efficiency than that of a 1 cM coupling-phase linkage. Marker-assisted selection do not need inoculation of pests, so that it avoids the errors caused by failed infection, incomplete penetrance of the resistance and variability of aggressiveness.

In addition, breeding for resistance can be carried out where inoculations of healthy plants in the field are not allowed for safety reasons. The susceptibility to Fenthion insecticide shown by tomato seedlings on detached leaves that carry the *Pto* gene of resistance to *P. syringae* pv. *tomato* is used as an indirect indicator to select for resistance to this bacteria. The isozyme marker *Aps-1* has been used commercially for many years as a substitute for screening with nematodes to select for the *Mi* resistance gene in tomato. *Mi* genotypes can now be selected by a PCR-based marker that is more tightly linked to *Mi* than *Aps-1*.

Screening tests for resistance to multiple pests are sometimes of doubtful validity because infection by one pest may interfere with the infections by other pests. Markerassisted techniques avoid infection and can help to introduce several genes each resistant to a different pest. Marker-assisted selection also offers considerable potential to transfer polygenic (quantitative) resistance because markers have high heritability (h=1 for molecular markers) and direct selection of resistance genes is masked by environmental effects. In tomato, molecular markers have been discovered for oligogenic and for polygenic resistances. A solution to control pathogens that infect roots is to use resistant rootstocks. They are used for several greenhouse crops such as tomatoes, eggplants, melons, watermelons, cucumbers, carnations and roses. However, for roses, rootstock grafting is done to improve disease resistance and to change the vigour and longevity of the crop.

Strategies to Improve Durability

The vast majority of the resistant cultivars rely on the use of single, major genes and these have proved remarkably successful, even though severe breakdown of resistance occurs from time to time. Several strategies are proposed to reduce the risk of resistance breakdown when major genes of resistance are used. Multilines or cultivar mixtures are formed either by phenotypically similar lines, or cultivars that each contain a different single, race-specific gene of resistance. No examples of multilines or cultivar mixtures occur among the species usually grown in greenhouses. Gene deployment uses several cultivars each with a different gene of resistance and grown within a clearly defined area. If the pest produces a virulent race on the cultivar grown, another cultivar that carries another gene of resistance will be grown in the area from next year until a new virulent race breaks its resistance. The next cultivar grown will either be the first cultivar or a new one with resistance to the last virulent race. Gene deployment, as multilines, exploits the diversity of the host-plant population to stabilize the pest population and avoid the appearance of virulent races. To effectively use gene deployment all the growers of the area must use cultivars with the same resistance gene.

Pyramiding resistance genes involves the introduction into the same cultivar, of all, or as many as possible, of the genes of resistance for a pest. The rationale behind pyramiding is that the pest will need several mutations from avirulence to virulence to overcome the resistance and that the probability of two or more successive mutations is extremely low because it is the product of the probability of each mutation. The gene *Pto* protects tomato against *P. syringae* pv. *tomato* race 0 and some resistant cultivars *Pto*/+ have been released. Stockinger and Walling found the novel genes of resistance *Pto-3* and *Pto-4* that can withstand races 0 and 1. According to Buonaurio, pyramiding *Pto, Pto-3,* and *Pto-4,* in one cultivar may provide the optimum solution for this disease control. Integrated pest management aims to keep the pest population continuously at a low level. Because the probability that new races of the pest will emerge is proportional to the population level of the pest, integrated pest management will reduce the possibility that a new virulent race will develop, and, consequently, the durability of race-specific resistance may increase.

6

Management of Plant-parasitic Nematodes

INTRODUCTION

Nematodes are roundworms, and those that attack plants are microscopic. The damage they cause to plants is often subtle and is easily confused with nutrient problems. Although hundreds of different kinds of nematodes may infect plants, less than a dozen are economically serious root-feeding pathogens in Hawaii, and only one genus here causes significant damage by feeding on foliage. If the numbers of harmful nematodes are large, plant growth is adversely affected. The primary groups of nematodes that cause problems in Hawaii are the root-knot, cyst, burrowing, lesion, foliar, and reniform nematodes. Others, such as the spiral, pin, and lance nematodes, may occur in abundance but usually do not cause sufficient damage to warrant concern. All of these nematodes, whether root-feeders or foliage-feeders, have a spear-like mouthpart that punctures plant cells so that the nematode can feed on the cell contents.

Feeding by some nematodes causes abnormal plant growth, and consequently the plants can be stunted, discoloured, or both. The environmental conditions of Hawaii are ideal for maximizing nematode damage. Nematodes thrive at the temperatures under which most crops and landscape plants are grown in Hawaii, especially where frequent rainfall or irrigation keeps the soil moist. Control of nematodes can be optimized by basingmanagement decisions on the relationship between nematode biology and plant response. Lacking a well informed management plan, arbitrary selection of control practices can be costly and ineffective.

PARASITIC NEMATODS

The name "nematode" comes from the Greek words: *nema,* which means "thread", and *toid,* which means "form". Nematodes are usually vermiform, long and slender, but some species are swollen. Most people know them as roundworms because their cross-section is round. They range in size from 0.08mm to 7m. Nematodes are a tube within a tube. The outside tube is the body wall which consisits of muscle layers that are used as a protective covering. The inside tube is the digestive system. The esophagus is the front of the digestive tube.

FEATURES OF NEMATODES

Aquatic, triploblastic, unsegmented, bilaterally symmetrical, roundworms, that are transparent, colorless, usually bisexual, and worm-shaped (vermiform), although some can become swollen (pyroform). Only an estimated 3 per cent of all nematode species have been studied.

HOME OF NEMATODS

Nematodes exist almost everywhere, including soil, plant and animal surfaces and interiors, decomposing life forms, and fresh and salt water and their subtending sediments. Nematodes can be grouped according to what they eat. The different groups are fungal feeders, bacterial feeders, predators, animal parasites, algal feeders,omnivores, and plant parasites.

MODE OF INFECTION

Plant-parasitic nematodes pierce plant cell walls with their stylet, which is formed by some of the mouth and esophagus parts. Then they pump up the plant cell just into their digestive system. These nematodes only feed on a certain part of the plant. Most feed on the roots. Some of kinds of nematodes feed only on the outer tissue of the root, others penetrate more deeply, and some completely enter the host. There are different ways they harm the plants. Some induce their hosts to produce nutrients which the nematode can survive on, or enlarged structures in which the nematodes live, or both. Some produce metabolites which kill host tissue. These activities all result in energy being removed from plants to support nematodes. Some plant-parasitic nematodes can cause severe damage that can result in unmarketablity, yield decrease, or even total crop failure.

SOME PLANT-PARASITIC NEMATODES ARE HARMFUL TO OREGON AGRICULTURE

Root-lesion nematodes (*Pratylenchus*) rip a hole in the sides of root cells and crawl inside. They move through the root, piercing, sucking, and leaving behind a trail of both cell-killing metabolites and eggs. Cell death results in brown lesions on the roots. Lesions begin on one side, but may encircle a root and thereby girdle it. The overall effect is a weak, shallow root system with a lot of dead areas. Root-knot nematodes (*Meloidogyne*) enter roots as juveniles, select a feeding site of three to eight cells, and swell up in their chosen spot as they progress towards adulthood. They introduce hormone-like substances into the plant cells, causing the plant to swell in the area producing galls, or root knots. Males regain their slender profiles and leave the root at adulthood, but the fattened adult females remain inside.

They exude eggs into the soil, and after the juveniles hatch, the cycle begins again. Roots infested with root-knot nematodes usually have visible galls and may exhibit excessive branching. Parasitized plants may be weak and stunted. Root systems may be deformed, and underground organs such as potato tubers and carrot taproots may be damaged and unmarketable.

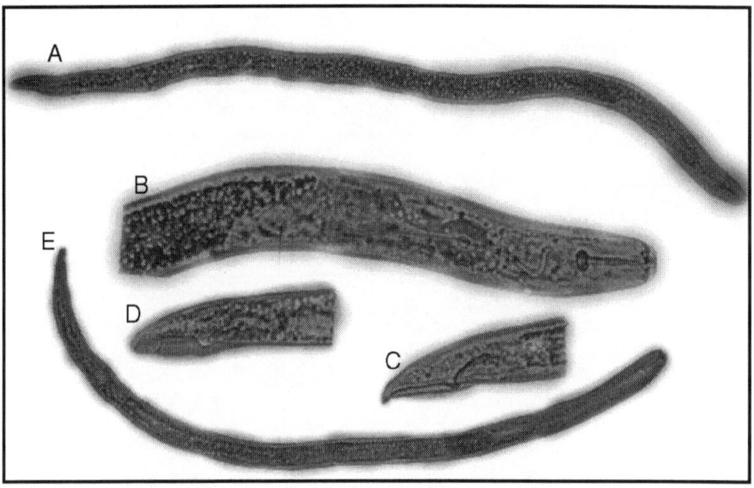

Fig. Vermiform (a) Adult Female Lesion Nematode, (b) anterior Region Showing Lobed Stylet, (c) Posterior of Female, (d) Posterior of Male Showing Spicule and Caudal Alae, (e) adult male lesion nematode

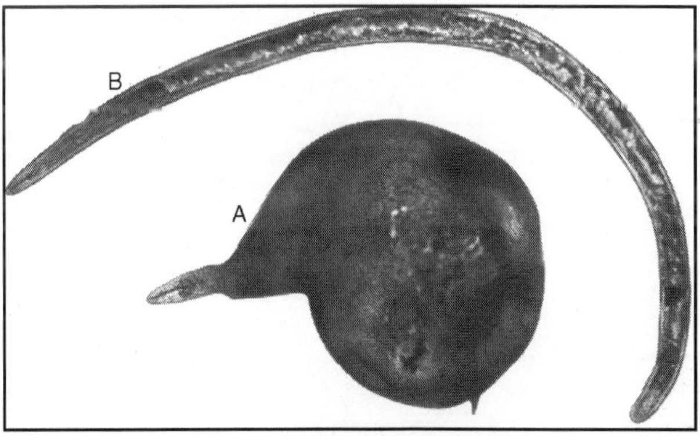

Fig. (A) Pyroform Female Root-knot Nematode, (B) Vermiform Male Root-knot Nematode

Pin nematodes (*Paratylenchus*) pierce root cells from the soil outside of the plant. Low numbers may appear inconsequential, but in high enough numbers, pin nematodes can damage crops. Ring nematodes (*Criconemella*) behave similarly. Little is known about ring nematode damage on many crop plants. Stubby-root nematodes (*Trichodorus* and *Paratrichodorus*) are also external soil-dwelling parasites. Besides piercing root cells and sucking out the contents, they can vector viruses such as the Tobacco Rattle Virus, which causes corky ringspot disease in potato tubers. Other virus-vectoring plant-parasitic nematodes are *Xiphinema,* daggar nematode, which can vector Tomato Ringspot Virus, a disease of many fruit crops, and *Longidorus* ("mint nematode").

Nematode Organ Systems

Skeletal - exoskeleton (triple layered cuticle)
Secretory/excretory -
Nervous - nerve ring and longitudinal muscles
Digestive - stoma, esophagus, intestine,
Reproductive -
NO circulatory or respiratory systems - exchange of gasses through cuticle.

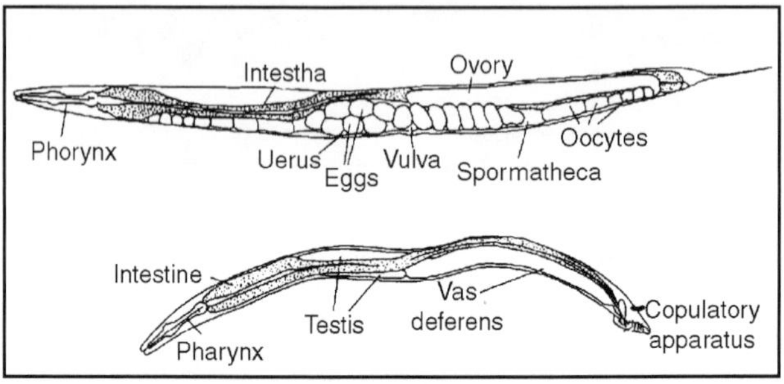

Fig. (Top) Hermaphrodite, (Bottom) Male

GENERAL PLANT PARASITIC NEMATODE LIFE CYCLE

5 Stages-4 Molts
Egg - J1 - M1 - J2 - (Hatch) - M2 - J3- M3 - J4 - M4 - Adult
In all plant parasitic nematodes the first molt occurs in the egg.
The emerging second-stage juvenile (J2) is the most common infective stage.

Length of life cycle of plant parasitic nematodes can be anywhere from 20-40 days (on average 25 days at 22°C).

NEMATODE MODES OF REPRODUCTION

Amphimixis - cross fertilization (males and females)
Parthenogenesis- nonsexual, males not required for reproduction(clonal)
Hermaphrodism - male and female organs in one individual(some change sexes)

Reproductive Potential (Egg production)

Anquina 25,000 eggs (2,000/wk)
Meloidogyne 500-1500 eggs
C. elegans 250-350 eggs
Ascaris 2 million eggs/day

NEMATODE GROUPS

1. Marine, Freshwater, Soil
2. Animal Parasites
A. Vertebrate
B. Invertebrate
3. Plant Parasites
Marine, Freshwater, Soil (Trophic Groups)
Microbivores ("free-living")
Fungivores
Omnivores (feed on multiple sources)
Predators (on nematodes, protozoa,etc.)

PLANT PARASITIC NEMATODES

PLANT PARASITES

197 Genera, 4300 species of phytoparasites Obligate parasites (cannot culture axenically, but can culture monoxenically)

EVOLUTIONARY ADAPTATIONS:

Plant Parasitic Nematodes have protrusible, hollow stylet, and morphological and physiological changes to the esophagus. All plant parasitic nematodes have a protrusible, hollow stylet, BUT not all stylet bearing nematodes are plant parasites.

MOST DAMAGING PLANT PARASITIC NEMATODES

Root-knot Nematodes

- Root knot nematodes cause more damage worldwide than any other genus
- some species can parasitize more than 2,000 plant species
- form complex feeding sites in root called "giant cells"
- galls are formed on root
- reproduce by parthenogenesis
- eggs laid outside the female body in a gelatinous matrix sac that often protrudes to the surface of the root and can be seen on the root surface as a sign
- 500-1000 eggs laid per female

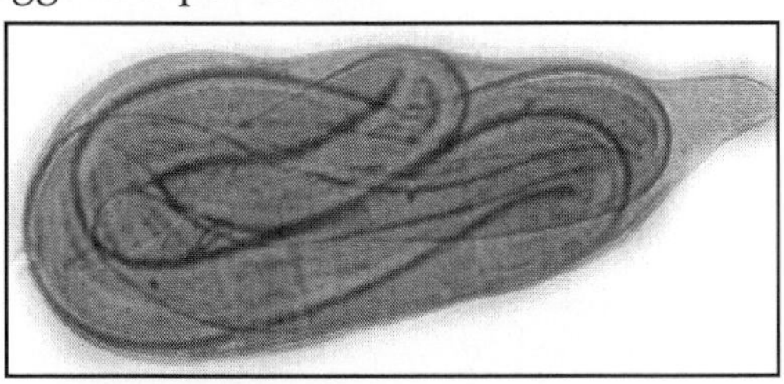

Fig. First Juvenile Molt Inside Egg.

At the establishment of a feeding site several dramatic physiological changes occur in both the host plant and the developing nematode including:

- Secretion of esophageal gland products possibly including hormones and enzymes required for feeding site formation and maintenance
- Degeneration of longitudinal muscles of J2 rendering nematode immobile (sedentary)
- Basal zone of the cuticle breaks down allowing for swelling during feeding
- Males third molt is a true metamorphosis seen in greater numbers under adverse environmental conditions by female sex differentiation (reversal)
- Undifferentiated vascular cells are targeted for feeding site initiation, most likely these are xylem parenchyma cells
- Feeding cells increase in size up to 100 per cent (hypertrophy)
- Cell contents multiply including mitochondria and nuclei (40-100 per cell)
- Walls develop elaborate ingrowths and become thickened possibly to provide better absorptive abilities to meet the nutritional demands of the growing nematode
- Giant cells degenerate upon nematode death, suggesting required stimulation by nematode for maintenance

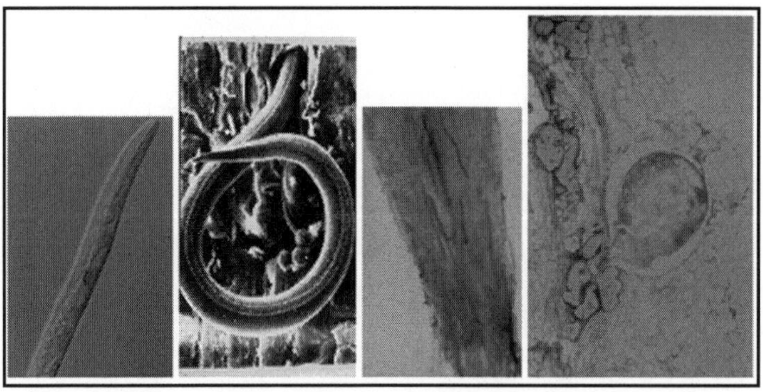

Fig. Infection Cycle of RKN.

- J2 free in soil,
- J2 penetrating root,
- J2s migrating to the vascular cylinder
- Adult female with 'giant cells' surrounding her head

Gall formation is a visible sign of root-knot nematode infection resulting from the hypertrophy and hyperplasia indicative of feeding site formation within the roots. The International Meloidogyne Project was conducted to assess the worldwide impact of this genus on agriculture and was based here at NCSU.

Some of the results of this project include:

- The North Carolina Differential Host Range Test

- Frequencies of the most common species:
  - *M. incognita* (Southern root knot nematode) 54 per cent
  - *M. javanica* (Javanese root knot nematode) 30 per cent
  - *M. hapla* (Northern root knot nematode) 7 per cent
  - *M. arenaria* (peanut root knot nematode) 7 per cent

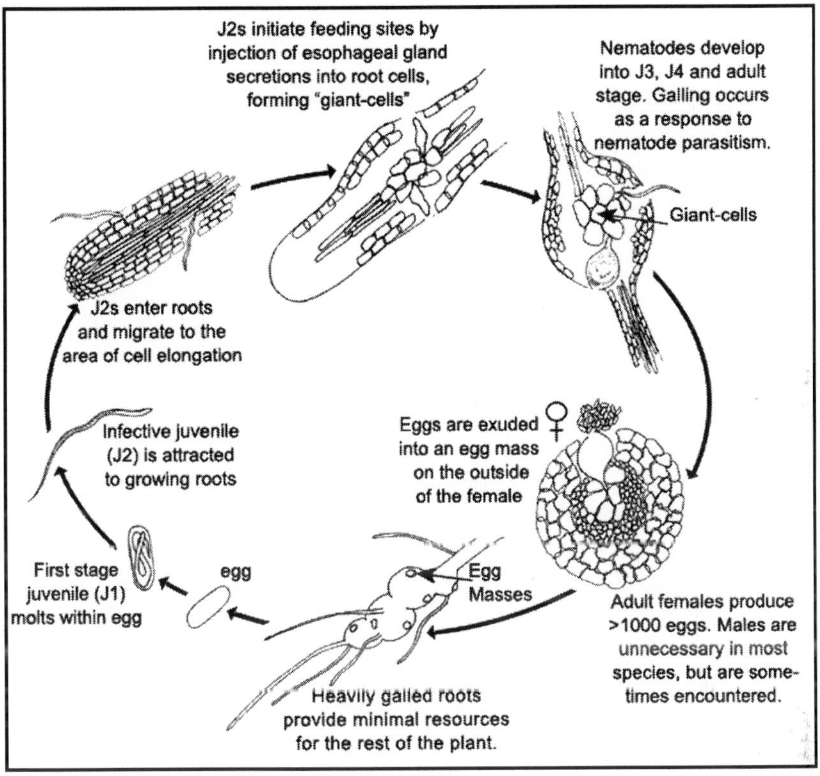

Fig. Root-knot Nematode Disease Cycle

The Cyst Nematode

Most Important Species

- *H. glycines* (soybean cyst - #1 pathogen of soybean)
- *H. schachtii* (sugar-beet cyst nematode; 218 host species)
- *H. avenae* (cereal cyst nematode)
- *H. trifolii* (clover cyst nematode)

General Characteristics

- Limited host-range.
- Reproduce by amphimixis.
- Form complex feeding sites called "syncytia".
- Cysts are lemon-shaped.

Life Cycle

- J2 cuts slit in egg shell with stylet(no enzymes involved)
- J2 migrates to root zone of elongation and penetrates to vascular cylinder
- Feeding site (syncytium) is initiated and J2 molts to sedentary J3
- Adult female produces 200-600 eggs that mature in the cyst
- Some species produce an egg mass for spontaneous hatching
- Eggs retained in cysts overwinter or survive up to 30 years(*G. pallida*)
- Hatching is often stimulated by hatching factors (root exhudates).

Hots Parasite Interactions

Fig. Left to Right: Broken Cyst with Eggs, Cysts on Roots, Juveniles Hatching from Cysts.

- Intracellular migration of J2 causes necrosis mechanically and with cellulases
- Syncitia are formed in pericyle and endodermal cells by dissolution of cell walls
- This syncytial initial is formed by the excretion of esophageal compounds
- Multinucleate cells result from the coalescence of neighboring cells
- Nuclei enlarge and cytoplasm becomes granulated
- The syncitium shows increased metabolism in response to the nematode which is acting like a nutrient sink in the infected roots
- Feeding tubes form from nematode and plant secretions in the syncitium at the distal tip of the stylet, but are not enveloped by endoplasmic reticulum as in root-knot feeding tubes
- A feeding plug is also formed at the point of cell wall penetration
- No galling due to the lack of hyperplasia and hypertrophy
- Degradation of the syncitium results in considerably more necrosis than root-knot

RELATIONSHIP BETWEEN NEMATODE BIOLOGY AND CONTROL

Management of nematodes must focus on reducing nematode numbers to levels below the damage threshold. In annual crops, the higher the nematode population numbers at the time of planting, the lower the yield. In perennial crops, the relationship between plant growth and nematode populations is more complex. The initial Plant-parasitic Nematodes and Their Management numbers of nematodes are still important because they determine the early growth potential of the plant. However, even if numbers are low at planting, nematode populations will eventually increase and ultimately damage perennial plants. Thus, not only must the initial populations at the time of planting be low, but also the populations must be kept at a low level if the plant or crop is to remain vigourous. Management of nematodes in the tropical and subtropical environments of Hawaii is a challenge. There are a few control measures that are effective, and these must be used under conditions in which they will work.

For effective management of nematodes, the critical steps are:

- Accurate diagnosis, and
- Proper selection of the most effective control method.

THE IMPORTANCE OF ACCURATE ASSAY

Accurate nematode diagnosis through a soil and/or root assay is the key to developing a successful management programme. Proper sampling is necessary to obtain a reliable and accurate estimate of the nematode population. Sampling before planting ("preplant" sampling) is one relatively reliable procedure for predicting the potential of nematode damage. For diagnostic purposes, samplesfrom a crop field or a group of plants can be collected any time after the plants have been growing actively for at least 2–3 weeks. Reliability of sampling is increased by sampling again. For preplant sampling, the repeat sampling can be done immediately. During the cropping period, the second sampling should be done a few weeks after the first.

Collecting Samples for Nematode Assay

The proper sampling method is based on the nematode and its location in the soil profile, which depends on the crop. With turf, for example, the nematodes occur primarily in the top 4 inches (10 cm) of soil, whereas with coffee and pineapple, the highest numbers of nematodes occur from 8 to 20 inches (20–50 cm) deep in the soil. Both large fields and small plantings such as home gardens should be sampled in a systematic, zigzag pattern. This sampling should consist of at least 15–20 cores or shovelfuls of soil composited into one sample per garden or 2–3-acre (3D4–11D4-hectare) section of a field.Research has shown that sampling in a pattern that systematically attempts to represent the entire area being sampled will produce a more accurate estimation of the actual numbers of nematodes than sampling at random. Single plants (such as a tree) should be sampled. Beneath the leaf drip line. The number of cores or shovelfuls of soil taken around a tree will depend on the size of the tree canopy. For a small tree (<10 ft [3 m] diameter canopy), collect 6–8 cores of soil and composite them into one sample. For larger trees, sample 10–12locations and composite them.

Handling the Nematode Sample

The composited soil should be gently mixed before withdrawing a subsample of about 1 pint (1/2litre). Place the subsample into a plastic bag and seal it to prevent moisture loss. Label the sample with your name, address, name of plant (including variety), and date. This sample must be handled gently and kept out of the sun, preferably in an insulated cooler or an area that is about 70°F(21°C). Send the sample to a processing laboratory such as the CTAHR Agricultural Diagnostic Service Center in a cardboard box or styrofoam container.

CONTROL OF NEMATODES

Nematode management should be multifaceted. Since eliminating nematodes is not possible, the goal is to manage their population, reducing their numbers below damaging levels. Common management methods used include planting resistant crop varieties, rotating crops, incorporating soil amendments, and applying pesticides. In some cases, soil solarization also may be practical.

CONTROL METHODS NOT INVOLVING PESTICIDES

Use of resistant plant cultivars is limited because there are only a few and their nematode resistance is very specific. Because resistance is specific, accurate identification of the nematode species and race is necessary before the proper cultivar can be selected. Crop resistance is ideally combined with a long-term crop rotation schedule and the best management practices available to favour vigourous and healthy plant growth. Crop rotation involves growing a crop that is not a host for the nematode present before growing a crop that is susceptible. The non-host or immune crop will cause nematode numbers in the soil to decline, giving the subsequent host crop a chance to establish a good root system. The success of this method depends on growing the non-host crop long enough to reduce the nematode numbers.

The rotation crop must be selected carefully because some nematodes (such as root-knot, reniform, and burrowing nematodes) have very wide host ranges.Also, crop rotation is difficult to use with most perennial crops. Variations on the crop rotation concept include fallowing, intercropping, and green manuring. Keeping the soil free of plants (fallow) deprives plant-parasitic nematodes of a host, which, over time, reduces their populations.

(Related to this concept is practicing good weed control, particularly when the weeds are hosts for the nematode.) Intercropping with plants that either are not good nematode hosts or are antagonistic to the nematodes also reduces nematode numbers. Green manuring, tilling under a crop that grows rapidly and produces a large quantity of biomass, adds organic matter and, depending on the green manure crop used, may add substances that repel or kill nematodes.

Another non-chemical approach to controlling nematodes is biological control—using other organisms against the pest organism. A high level of natural biological control is ordinarily present in the soil. This natural control probably keeps the nematode populations a 10–20 per cent of what they would be in its absence. Nevertheless, the level of natural control is seldom adequate to prevent plant damage from nematodes. The strategy of inoculating soils with biological control organisms to increase or supplement the control organisms naturally present has proven to be unrealistic and is not recommended until more predictable inoculants are developed.

ROOT-KNOT NEMATODE

Root-knot nematodes belong to the genus *Meloidogyne*. The two most common species in Hawaii are *M. incognita* (southern root-knot) and *M. javanica* (Javanese root-knot); other species are present but occur less frequently. The root-knot nematodes feed and mature inside the roots of plants. Their feeding induces abnormal enlargements of the root called galls. The root-knot nematode does not survive very long without a host plant, except in very low numbers and probably in the egg stage. *Diagnosis:* Root galls are the primary symptom of root-knot nematodes. Species identification requires laboratory assay.

RENIFORM NEMATODE

The reniform nematode, *Rotylenchulus reniformis*, is widespread throughout Hawaii and feeds on many cultivated and non-cultivated plants. The juvenile stages and males live in the soil and do not feed. The adult female is swollen and is the only parasitic stage of this nematode's life cycle. The female inserts her head and neck into the root, leaving her body outside of the root. The reniform nematode survives in the soil as eggs and coiled juveniles. This nematode causes root rotting and reduced uptake of water and soil nutrients. The symptoms are general lack of vigour and discolouration of foliage, and (or) stunted plants. *Diagnosis:* The reniform nematode can be accurately diagnosed only through laboratory assay of a soil sample.

Burrowing and Lesion Nematodes

Adult burrowing nematodes (*Radopholus similis*) and lesion nematodes (*Pratylenchus* spp.) cause root rot. These worm-shaped nematodes are migratory, living most of their life, including the hatching and feeding stages, inside the roots and sometimes the lower stem of their host plants. They can move through the soil from one root to another during their life cycle. *Diagnosis:* The root rot caused by this nematode is not characteristic for diagnosis. It is necessary to have the soil and roots assayed to determine the numbers and kinds of species present; root assays are the most reliable.

Sugar-beet Cyst Nematode

Sugar-beet cyst nematode (*Heterodera schachtii*), is primarily a problem on cabbage, broccoli, and cauliflower in Hawaii in the Kula, Maui, area. This nematode penetrates the root, and the female enlarges as it matures to become a white, lemon-shaped structure that breaks through the root surface at maturity. When the female dies, her body turns brown. Eggs survive inside the dead female body (called a cyst) for many years.

Diagnosis

Direct observation of the organism with a magnifying glass is helpful, because the white cysts are about the size of the period at the end of this sentence. Root aphids are also white, so some experience may be needed to differentiate between these two white structures clinging to the roots. Confirmation by a diagnostic laboratory is recommended.

Foliar Nematode

Aphelenchoides besseyi, A. ritzema-bosi, and *A.fragariae* feed inside foliage. The entire nematode life cycle is completed in the leaves. Plants can be stunted with deformed, discoloured, or dying leaf tissue; "dieback" can also occur.

Diagnosis

Accurate identification requires laboratory assay of leaf-tissue samples. Nematodes a more realistic strategy for biological control of nematodes is to incorporate soil amendments such as manures (particularly chicken manure) and compost. Such additions of organic matter contribute to biological activity in the soil and enhance the natural activity of organisms antagonistic to nematodes.

Solarization, the heating of soil by using clear plastic tarps to increase and trap the sun's heat, can be an effective means of controlling nematodes in the soil. The soil needs to be moist, well tilled, and heated to at least 140°F (60°C) for several days, preferably several weeks. This method can be practical for home gardens, but it should be done during the hot months of mid-summer.

Similarly, other heat and steam-based pasteurization methods can be used to prepare potting soil. Healthy plants grown in nematode-free media have a better chance to survive after being transplanted to the soil.

Pesticides (Nematicides)

Nematicides are sometimes used in agriculture, but few of them are available in Hawaii. Most nematicides are highly toxic synthetic pesticides commercially available only to commercial growers. These products can be used only on particular crops, and they usually must be purchased and applied by a licensed pesticide applicator. Two types of nematicides are fumigants and non-fumigants. Fumigant nematicides are usually more effective, but non-fumigant nematicides can also be used effectively. Fumigant nematicides such as metam sodium.

Suggestions for Managing Nematodes in the Home Garden

- Susceptible plants can be grown in containers with a nematode-free soil or growth media; keep the containers off the ground.
- Sow seedlings for transplanting in clean media in containers kept off the ground.
- Check the roots of container-grown nursery stock for nematode galls before transplanting them.
- Add large amounts of organic matter to the soil.Keep weeds controlled at all times.
- As soon as plants are not needed, pull them up or till them in to stop their roots from hosting nematodes.
- Infected plant roots systems should be destroyed and not composted.
- Rotate susceptible plants with resistant orimmune plants.and 1,3-dichloropropene are applied before planting.Some non-fumigant nematicides such as Nemacur®,Mocap®, or Vydate® are moderately effective and can be used both pre- and post-planting. Some "natural" products claiming to provide control of nematodes have been developed from biological sources. The apparent recovery of plants resulting from using these products can sometimes be due to growth enhancement by plant nutrients contained in the formulations. As with all pesticides, nematicide use is subject to state and federal regulations. The pesticide label must specifically allow its use in the crop or situation, and label directions must be followed. *Donald P. Schmitt and Brent S. SipesDepartment of Plant Pathology*

A Note to Commercial Growers

For assistance with diagnosis of crop problems that may be due to nematodes, contact your local office of the Cooperative Extension Service. Soil assay for nematodes and identification of the nematode species is necessary to develop a management plan. Plant-parasitic nematodes must be addressed in crop production and integrated pest management (IPM) systems if agriculture is to meet the world demands for increasing food and fibre production. On a worldwide basis, annual crop losses due to nematode damage have been estimated to average 12.3 per cent, amounting to some US$77 million annually. The data available indicate that similar losses occur in the Near East. For example, estimated losses for vegetable crops due to nematode-related disease complexes in Egypt amounted to some 15 per cent in 1986, with losses for field crops ranging from 5 to 20 per cent. Nevertheless, the subtle nematode symptoms and signs are often confused with nutrient deficiencies and other maladies, resulting in nematodes being overlooked by agricultural scientists as well as growers.

The concept of combining compatible tactics for controlling nematodes predates that of IPM. In 1889, Atkinson discussed a range of nematode-management tactics that is surprisingly similar to those available today. Tactics discussed for controlling root-knot nematodes include: sterilization of the soil by starvation (including the use of non-host plants); the potential of trap crops; composts; nematicides; and soil amendments such as hardwood ash and potash. The early work of Cobb on sampling nematode communities provided a basis for the development of improved tactics and strategies essential for integrated pest management. The following statement was offered by Tyler in 1933: "A well-planned combination of practices will go much further for control of nematodes than any of the recommended treatments alone". Tyler also stated that the presence of nematodes (root-knot) in any soil can best be determined by examining the roots of susceptible plants which have been growing for at least three weeks in warm, moist conditions.

For attempted eradication of root-knot nematodes, the author suggested the following:

- Burning of residues two or three times if possible, each preceded by digging or ploughing.
- Dry fallows with frequent ploughing.
- One or two well-irrigated trap crops completely destroyed two or three weeks after sprouting.
- Moist fallows during warm weather, without weeds.
- Resistant crops in rotation, kept free of weeds.
- Repetition of steps 4 and 5.

Tyler also emphasized that root-knot nematodes enhance the susceptibility of crops to other diseases such as cotton wilt, black shank of tobacco and *Rhizoctonia* disease of groundnuts. Thus, early nematologists provided an exceptionally strong conceptual framework for the development of advanced integrated management systems.

The specific concept of IPM is a relatively recent development in pest control. The 1962 publication of *Silent Spring* by Rachel Carlson and the subsequent banning of DDT awakened United States agriculturists, especially entomologists, to the risks of pesticides and provided the impetus for the development of IPM, IPM consists of the development, use and deployment of pest-control strategies and tactics that result in favourable socio-economic and environmental consequences. Theoretically, IPM is a holistic systems approach to limit pest damage to tolerable levels through a combination of tactics and techniques, including parasites and predators, host resistance, cultural practices, environmental modification, and pesticides where appropriate. Unfortunately, the interest and progress of IPM programmes has not been balanced among the pest disciplines with the major emphasis often being on insects. Still, a number of national and international IPM programmes have resulted in many research and extension programmes becoming more interdisciplinary.

The relative youth of the science of nematology and the availability of effective, relatively inexpensive nematicides hampered the development of integrated management systems in this discipline. During the last two decades, however, much progress has been made in characterizing damage functions and thresholds for major nematode-crop combinations, the study of general population dynamics and simulation models for key nematodes, and the development of improved management tactics. In addition, considerable progress has been made in elucidating the roles of nematodes in disease/pest complexes, and their interactions with symbionts.

BIOLOGICAL CONTROL AGENT FOR NEMATODES

Too often, biological control agents have failed because they have been used before a basic knowledge of their ecology and biology has been established. The importance of such knowledge can be seen from work at Rothamsted aimed at the development of *V. chlamydosporium* as a biological control agent for root-knot nematodes. Stirling describes the design and methods used in the conduct of biological control experiments, which are more complex than those required to assess the efficacy of a nematicide. He identified five key aspects in setting up an experiment to evaluate a biological control agent and these are presented below in a slightly modified form.

- The test organism and any organic amendment should be applied at practical application rates; 0.1 per cent w/w soil is equivalent to 2.5 tonnes/ha and should represent a maximum dose. Tests should always be performed in a non-sterilized soil with a natural residual soil microflora.
- Appropriate treatments as well as an untreated control should be included if the organism is added with a substrate. These treatments should include the substrate alone, the organism alone, and the autoclaved colonized substrate. Too often, untreated controls are compared only with large applications of the organism and substrate and this does not allow separation of the effects of the agent from the effects of the substrate. In several tests reported in the literature,

application of the substrate alone has decreased nematode populations to the same extent as the substrate colonized by the agent, and there is no clear evidence of biological control.

- Population densities of the agent under test should be monitored to ensure that it has survived in soil throughout the period that activity against the nematode target is required. Such monitoring may require the development of selective media, which can be a difficult and time-consuming task.
- Nematode mortality caused by the organism under test should be measured to assess whether differences between nematode population densities in treated and untreated soil relate to the levels of kill caused by the agent. Infection levels are relatively straightforward to estimate for most parasites, but repeated sampling is required to determine total kills. The effects of agents which produce toxins or have indirect effects on nematodes through competition, the modification of root exudates or the colonization of feeding cells, can only be measured by assessing their impact on nematode development.
- The impact of the soil environment, host plant and nematode should be tested as these are likely to affect the efficacy of the biological control agent, and could account for the lack of activity of potential agents in specific test conditions.

In a review of the literature less than 15 per cent of experiments purporting to demonstrate biological control caused by *Paecilomyces lilacinus* satisfied the above criteria. Although this is an unsatisfactory situation that must be remedied, the difficulties in conducting carefully controlled and monitored experiments should not be underestimated.

Isolates of *V. chlamydosporium* differ markedly in their growth and sporulation *in vitro*, and in their virulence, saprophytic competitiveness and rhizosphere competence. Such differences between isolates of the same species of micro-organism are common and there is a need for simple laboratory-based screening methods to select the most promising isolates for further testing. *Verticillium chlamydosporium* isolates were collected from infested nematode females and eggs in suppressive soils around the world. Hence, they were isolated from the niche in which they would be required to be active as a biological control agent and from soils most likely to yield active isolates. Once isolated in pure culture using standard techniques, the *in vitro* growth requirements were determined.

Virulent isolates were selected by counting the number of nematode eggs parasitized after exposure to the fungus on agar in a standard test. Tests for proliferation in the rhizosphere and growth in soil were also used to select promising isolates. Although simple laboratory-based screens help eliminate many isolates that show insufficient activity to justify further testing, selected isolates will not necessarily be active in the field. As it may take 10 to 16 weeks to investigate adequately the performance of different isolates against cyst and root-knot nematodes in pot tests, relatively few can be screened. The method of mass culturing of *V. chlamydosporium* for experiments can have a marked effect on the subsequent survival and proliferation of the fungus in soil. Inoculum produced in shaken liquid cultures consists mostly of hyphae and conidia, which require an energy source to ensure proliferation in soil, whereas on solid media large numbers of chlamydospores are produced and these can be added to soil in aqueous suspension and rapidly establish the fungus.

The development of a semi-selective medium has enabled detailed studies to be made on changes in relative abundance of the fungus in soil and on roots. Some isolates of *V. chlamydosporium* may be extremely abundant in soil but unless they are capable of colonizing the rhizosphere they do not parasitize the eggs of root-knot nematodes. Growth in the rhizosphere differs markedly between plant species, *e.g.*, tomato, cabbage and maize roots support much growth, whereas sorghum, pepper and cotton are poor hosts. Colonization by the fungus is confined to the rhizosphere and rhizoplane and there is no spread into root tissue; no lesions have been observed on roots grown in soil treated with *V. chlamydosporium* and there have been no detrimental effects on the growth of a range of crop species.

A tenfold reduction (from 10^4 to 10^3 chlamydospores/g soil) in the amount of fungus applied to soil had no effect on the extent of colonization in the rhizosphere; the ability to proliferate on the root surface where the fungus is required to control nematodes is an important characteristic which may allow significant reductions in the amount of inoculum applied to soil.

The efficacy of *V. chlamydosporium* as a biological control agent for root-knot nematodes is affected by three key factors: the amount of fungus in the rhizosphere; the rate of development of eggs in the egg masses; and the size of the galls in which the female nematodes develop. In large galls female root-kn2ot nematodes may produce egg masses which remain within the gall and are not exposed to parasitism by *V. chlamydosporium,* which is confined to the rhizosphere. Hence, *V. chlamydosporium* is less effective in controlling root-knot nematodes in heavily infested soils and on highly susceptible crops because large galls are formed on the roots and many eggs escape parasitism.

Verticillium chlamydosporium is unlikely to be useful in these situations where a grower would normally apply a nematicide. Also, at temperatures above 25°C eggs may complete their embryonic development and hatch before the fungus has completely colonized the egg mass; at 30°C about 30 per cent of eggs of three root-knot species hatched and the second-stage juveniles escaped from the egg mass before the eggs were killed. These studies in pot tests, if supported by field experiments, help to define the conditions in which

V. chlamydosporium might be used successfully for control of root-knot nematodes. It is only from such detailed studies that the limitations and requirements of the fungus can be assessed.

The Extent of Colonization of the Rhizosphere by *Verticillium chlamydosporium* on Several Plants Grown on Soil Treated with 5 000 Chlamydospores/g Soil and the Control of *Meloidogyne incognita*

Verticillium chlamydosporium is not a replacement for nematicides but, despite its limitations, it may be a useful management tool. In most tests biological control agents have been applied to protect susceptible crops. However, application of the fungus to a relatively poor host for the nematode on which small galls are produced so that most egg masses are exposed in the rhizosphere might provide more effective control; soil population densities of the nematode would be reduced to non-damaging levels before a susceptible crop was planted.

Host plant	Rhizosphere colonization (cfu/cm² root)[1]	Control (%)
Tomato	229	58
Maize	216	71
Cabbage	199	68
Potato	88	79
Tobacco	43	40
Pepper	29	36
Sorghum	25	0
Wheat	15	5

Note:

[1] cfu are the colony-forming units which develop on the semi-selective medium.

Bridge recommends the following rotation for the management of root-knot nematodes: susceptible host - poor host - poor host - resistant or non-host -susceptible crop. Application of the fungus before the first or second poor host may permit the shortening of rotations, or the replacement of the resistant or non-host crop, without increasing nematode damage to the susceptible crops. Currently, at Rothamsted Experimental Station, United Kingdom, poor hosts for root-knot nematodes are being screened for their ability to support *V. chlamydosporium* in their rhizospheres. It is hoped that by combining the host plant and the biological control agent, more effective and consistent control can be achieved.

The Next 20 Years

The past 20 years have seen a significant increase in the number of scientists involved in research on the biological control of nematodes. Surveys and empirical tests are being replaced by quantitative experimentation and basic research on the modes of action, host specificity and epidemiology of selected organisms. Such basic information is essential for a realistic appraisal of the impact of molecular biology on the improvement of microbial agents and monitoring the spread and survival of released organisms, and for the development of rational strategies for control.

Current experience suggests that biological control agents will not replace the use of nematicides but, integrated with other control measures including chemicals, they could play an important role in the development of integrated control strategies in both developed and developing agriculture. The urgent need to reduce the dependence on nematicides should provide the necessary impetus for the considerable amount of research and development still required to ensure the successful use of such agents.

Pathology

Bailey and Gordon reported that *R. culicivorax* depleted host metabolites, reduced fat body and other host storage tissues while accumulating stored materials in their trophosomes, and thus inhibited the development of imaginal disks in the host mosquito. Hemolymph proteins were depleted sixfold in mosquitoes.

Mermithid parasites of larval diptera probably resemble *M. nigrescens* in obtaining dietary amino acids by stimulating the catabolism of proteins within the host fat body. In addition to the reduction of most protein fractions, significant decreases also occurred in hemolymph glucose in parasitized blackflies; however, blood trehalose concentrations were not affected. Glycogen reserves were shown to be similarly reduced in the fat body of mermithid- parasitized simuliids. Gordon *ET al.*found that parasitism did not significantly affect either the overall concentration of lipids or relative proportions of the lipid fraction in the hemolymph of mosquitoes. Mermithid parasitism of mosquitoes and simuliids caused almost complete degeneration of host fat body tissue. All storage metabolites, including glycogen, within the fat body may directly or indirectly be utilized by the developing nematode.

Responses are somewhat different in larger hosts with longer periods of parasitism. Protein in the fat body of grasshoppers parasitized by M. *nigrescens* was depleted after 2 weeks; corresponding reductions in proteins were not recorded from the hemolymph until I week later. The third week of infection represented a period of maintenance and reconstruction of fat body soluble proteins, but corresponding effects were not found for hemolymph proteins until the fourth week of infection.

Gordon and Webster found that the overall level of amino acids in the hemolymph of adult female grasshoppers was not affected by mermithid parasitism, but the content of amino acids within the fat body was significantly reduced. Parasitism by *M. nigrescens* resulted in glucose levels remaining low for the first 4 weeks of parasitism relative to trehalose, but no significant difference for glucose levels when compared to unparasitized controls. Trehalose levels dropped and remained low for much of the parasitic period. These data suggested that the nematode modifies its host's metabolism to favour production of small molecules such as glucose and amino acids from carbohydrate and protein reserves and that the host cell's ability to sequester nutrients is impaired during parasitism.

Craig and Webster found that ecdysone levels in locusts were unaffected by *M. nigrescens* and attributed the inhibition of molting in parasitized insects to depletion by the nematode of precursors required by the host for protein and cuticle synthesis. However, ovaries of parasitized locusts were unable to sequester vitellogenic proteins available within the hemolymph, suggesting endocrine dysfunction in the host.

Major modifications of the host's metabolism by mermithid parasitism are manifested in host tissue degeneration and retarded development including resorption or suppression of oocyte development. As a result, hosts are usually prevented from maturing to the adult stage or when they do they are generally unable to-, reproduce.

Hosts are occasionally able to prevent parasite development by either cellular or humoral responses. The most common cellular response is encapsulation. Poinar *et al.* reported that *Culex territans re*sponded to the development of parasitic juveniles of *R. culicivorax by* covering the 2- to 3-day-old parasites with blood cells. These nematodes failed to complete development. A more frequent condition, melanotic encapsulation, was reported in *Diabrotica* beetle larvae attacked by F. *leipsandra.* Host hemocytes lysed on the cuticle of the nematode soon after it entered the host, and within 6-8 hours an inner layer of melanin had formed around the nematode.

Similar responses were observed for larvae of *Psorophoraferox, Aedes triseriatus, and Anopheles quadrimaculatus* to the *R. culicivorax. How*ever, some 70 species of mosquitoes have not elicited this type of defence response to *R. culicivorax*. Recently, Gaugler *et al.* reported an encapsulation response when *Empidomermis* sp. entered larvae of *Aedes stimulans* if the nematodes were unable to migrate to the host's head and enter central nervous tissue.

After pupation, the nematodes dropped back into the hemocoel where they no longer elicited a host defence reaction. This response may be common in mermithids that enter the larval stage but fail to mature until the host reaches the adult stage. Also, Harlos *et al.* noted that *Culicimermis* sp. entered the nerve tissue of *Aedes vexans* larvae and later developed in adult hosts; Petersen (unpublished data) observed a similar response in *Ae. sollicitans* when parasitized by *P. culicis.*

Humoral responses, non-cellular components of the insect's hemolymph which have an adverse effect on nematode development, may account for the host resistance observed in *Aedes intrudens* and *Aedes provocans* to an

Empidomermis sp. when no discernible host response was observed, which was unlike the encapsulation response this nematode elicited in *Ae. stimulans*. Another type of humoral response may be that observed when blackflies were subjected to the mosquito mermithid *R. culicivorax*. The mermithid initiated development in early-instar; lackflies, and direct host reaction to this unnatural parasite was not observed; eventually, however, both nematode and host died.

Physical and behavioural responses also influence parasitism by mermithids. Host age, especially in aquatic insect hosts, has been shown to have a considerable effect on host susceptibility. Refractiveness because of age appears to be caused by the thicker cuticle in older larvae. In the laboratory, as a result of mass-rearing practices for *R. culicivorax,* where exposed but uninfected hosts were retained to supply the next generation, the host mosquitoes were found to have become measurably less susceptible to the nematode after some 100 generations. No evidence of humoral defence mechanisms was found and the mode of defence was not determined. In a similar buildup of refractiveness in *An. quadrimaculatus to S. Petersen,* it was reported that less susceptible hosts were more active and aggressive in attempting to remove attacking preparasites.

Natural Population Regulation

Mermithids generally have localized, discontiguous distribution. This is especially true for mermithids that develop in larval hosts. *Romanomermis culicivorax* was found in only seven pools out of several thousand examined in southwestern Louisiana. This is understandable because the hosts are killed prior to emergence from the habitat, thus greatly reducing the nematodes ability to disperse.

Generally, observations of mermithid parasitism are based on isolated collections and often give the impression of higher parasite activity than actually exists. In extended studies, levels of parasitism are usually found to vary greatly. In a 27-month survey of five habitats containing populations of *R. culicivorax,* mean parasitism of *Anopheles crucians* for individual habitats ranged from 8 to 42 per cent, with parasitism of individual populations as high as 80 per cent. Similar but lower levels of parasitism were observed for several other mosquito species.

Similarly, Phelps and De Foliart observed parasitism of blackflies, *Simulium vittatum,* in four Wisconsin streams over a 1-year period. Parasitism typically ranged from 10 to 90 per cent and the authors suggested that localized parasite populations probably attained levels of parasitism sufficient to virtually eliminate some host populations. Similar population reductions of blackflies were reported by Welch and Rubtsov.

Parasitism of adult aquatic insects usually occurs over a wider geographical area and often reaches very high levels but fluctuates greatly over time. Steiner observed 80 per cent parasitism of adult *Ae. vexans in* British

Columbia in 1920, but only 20 per cent the following year. Trpis *et al.* reported 100 per cent parasitism of the same host in the same locality in 1967. Similar observations were made on parasitism of *Ae. sollicitans* adults by *P. culicis* in southwestern Lousiana. During an entire year only collections made in January failed to produce infected hosts; mean monthly parasitism ranged from 0 to 48 per cent with individual populations reaching 96 per cent. During a period of high parasitism, a survey was made along 80 km of coastline and revealed parasitism ranging from 48 to 94 per cent for seven populations.

Similar isolated high levels of parasitism are frequently observed in terrestrial insects. Christie reported a heavy outbreak of grasshoppers in Wisconsin in 1923-1925 with a correspondingly heavy infestation *of M. nigrescens.* The author concluded that the parasites were an important factor in terminating the grasshopper problem. Similarly, Mongkolkiti and Hosford reported that a population of the grass*hopper Hesperotettix viridis pratensis* was totally destroyed prior to egg lying by *M. nigrescens.* Other species of grasshoppers in this population were only partially affected or not affected by the mermithid.

Though reports of natural insect control are impressive and suggest the potential of mermithids as biological control agents, these parasites generally do not influence significant control over host populations. In a review of blackfly-mermithid literature, Molloy reported that parasitism in most blackfly populations is moderate, ranging from 3 to 15 per cent, and is perennial with only rare and highly localized epizootics. This pattern probably holds true for most mermithid-host associations.

Safety

As previously mentioned mermithids are generally specific to one or a few species of insects and are rarely found in other insects. As a result, little concern has been shown for potential danger of these parasites to mammals. To date, only limited mammalian safety studies have been made. Ignoffo *et al* reported that when suckling and adult mice and adult rats were subjected to either *per os,* intranasal, intraperitoneal, or dermal challenge of *R. culicivorax,* their body weight gain and histologies were identical to those of untreated animals. Immunodepressed rats also were not susceptible. Similar findings were reported for *R. iyengari in* India. Similarly, studies in the People's Republic of China showed that no pathology was apparent in dermally challenged suckling mice, in orally exposed adult rats, or in three species of fish when exposed to *R. jingdeensis.*

Mermithids have occasionally been reported as accidental parasites of man. Poinar reviewed all such cases in the literature and concluded that in most cases the data were insufficient to ascertain if human parasitism actually occurred. He further concluded that the question of accidental human infection by mermithids cannot be definitely answered at this time and should only be accepted as fact when proven experimentally or when parasites are found developing *in situ* in the human organism. Further, the Environmental Protection Agency has exempted mermithids from safety regulations.

Mass Propagation

Mass production systems at economical costs are essential if a given biological control agent is to be effectively used. Mermithids take in their nutritional, needs directly through the cuticle, making them especially difficult to propagate by *in vitro* methods. Therefore, pest control attempts have been limited to those mermithid-host systems that are conducive to *in vivo* rearing.

In Vivo Propagation

The *in vivo* rearing of mermithids of terrestrial insects is especially difficult because of the protracted life cycles of many of the hosts. Though numerous mermithid species have been encountered, laboratory culture has been limited to *M. nigrescens and F. leipsandra.* Laboratory infections are relatively easy with *M. nigrescens.* Gordon and Webster collected adult females of *M. nigrescens* from the field, allowed the females to oviposit, concentrated the eggs on moist filter paper, and stored the eggs at 5°C until needed. To infect hosts, eggs were transferred onto small pieces of grass previously coated with an adhesive material. The grass was readily eaten by host grasshoppers. This method permitted extensive research on the physiological effects of mermithid parasitism on a host system. The long developmental period of the nematodes during both the parasitic and free- living stages (I year) has prevented meaningful use of this nematode in field trials.

Creighton and Fassuliotis were the first to give a detailed report of the laboratory culture of a mermithid of terrestrial insects. Preparasitic juveniles of *F. leipsandra* and the larvae of the cucumber beetle, *Diabrotica blateata,* were placed in holes in soil in clay pots. The pots were sealed and held for parasite emergence. Ratios of 1000 nematodes per 100 host larvae were sufficient for maximum recovery of postparasitic nematodes. The method has the potential of producing between 80,000 and 240,000 nematode eggs per experimental unit. From this basic procedure, Creighton and Fassuliotis developed an *in vivo* mass rearing system for *F. leipsandra.* As eggs were laid, they were concentrated, sterilized by immersion for 10 minutes in 0.28 per cent sodium hypochlorite, and rinsed.

The eggs were transferred to a Baermann funnel and allowed to mature and hatch. First instar larvae of the banded cucumber beetle were exposed to the preparasites in clay pots as previously described. Twenty-four pots were placed in a plastic box and held 4-5 days at 24-27°C. The contents of the clay pots were then transferred to porous baskets. The baskets were placed at one end of a larger container with sand at the other end. As the host larvae matured, they migrated to the sand and pupated. After about 10 days the nematodes began to emerge from their hosts. The postparasites were then transferred to a container of aerated water until the nematodes molted to the adult stage and began to deposit eggs (7-10 days). Each larger container (37 x 27 x 10 cm) produced an average yield of 1.9 X 106 nematode eggs. The system can produce about 5 x 106 eggs week- I at a cost of $0.40 (US$) per I 000 eggs including supplies and labour for producing the host insects.

Although mermithids are relatively common in aquatic insects, especially in the Simuliidae, Chironomidae, Ceratopogonidae, and Culicidae, many of these host species cannot be successfully maintained in laboratory colonies or cannot be maintained in large numbers. Also many mermithid species undergo egg diapause or have an asynchronous hatch which prevents their mass production. As a result, few aquatic mermithid species have been studied extensively. The midge parasite, *Hydromermis conopophaga,* has been readily maintained in colony because the host, *Tanytarsus* midges, can be easily maintained. However, *H. conopophaga* has never been mass produced or released in the field. Similarly, the mermithid *Limnomermis rosea,* a parasite of chironomid larvae, has been reared through several generations but has not been released in the field.

Muspratt was the first person to develop, maintain, and report a laboratory culture of a mermithid nematode. He collected *0. muspratti (Agamomermis)* from naturally infected host mosquito larvae and placed the nematodes in containers filled with moist sandy soil. The soil was allowed to dry out over a period of several months. After 11-12 months some of the soil was placed in a container of water and first instar mosquitoes added. Seventy to 80 per cent of the host larvae contained nematodes. The original work by Muspratt though very basic provided the direction for future work with mosquito mermithids. Subsequent studies with 0. *muspratti* have refined the rearing technique and defined biological limitations. A mass production system for *0. muspratti has* failed to develop, even though a suitable laboratory host system is available, because the eggs of this species failed to hatch at the same time although they appeared to mature. Because of the non-synchronous hatch, cultures were observed to produce infective stage nematodes after periodic floodings for over 5 years (Petersen, 1981).

The mermithid, *S. peterseni,* a parasite of *Anopheles* mosquitoes, has been maintained in colony for over 10 years. Though this nematode appears to be an infective parasite, and is easily maintained in laboratory colony, economical mass production systems have been difficult to achieve because the only hosts are *Anopheles mos*quitoes, which are less suitable for mass rearing systems than are certain *Aedes and Culex* species.

The most efficient and effective *in vivo* rearing system for mermithids is that developed for *R. culicivorax*. The primary techniques were developed in 1971. This initial system called for the exposure of 20,000 first-instar *Cx. quinquefasciatus* larvae in 136 x 52 x 5 cm galvanized 'Lrays to preparasites at a parasite-host ratio of 12: 1. The hosts were then fed a regimented diet for 7 days. After this period the culture trays were drained and the mosquitoes concentrated. Pupae of uninfected mosquitoes were separated using a chilled water procedure, and the infected mosquitoes were placed in 36 x 25 x 10 cm trays and held for nematode emergence.

Postparasitic nematodes were then washed and 10-15 g placed in paraffin-coated aluminum pans (22 x 33 x 5 cm) which contained clean, coarse, sterile sand and water. After 3 weeks the free water was removed and the cultures stored an additional 4-15 weeks before use. When preparasitic nematodes were needed the cultures were flooded with chlorine-free water to stimulate nematode hatch. Nematode cultures averaged 1.91 million preparasites (45-60 per cent of total yield) if they were flooded for the first time when they were 11- 19 weeks old. Total yields were highest (5.32 million preparasites) when cultures were flooded for the first time when 8-10 weeks old and then at 3-4 week intervals thereafter. It has since been determined that yields can be increased by having a density of 24 postparasites CM-2 in the culture trays. Yields of preparasites were tripled by simply setting up three cultures (22 x 33 x 5 cm), each containing 5 g of nematodes, instead of the usual method of one culture containing 15 g of postparasites. The longevity of cultures could be extended for up to 9 weeks if the cultures were allowed to mature at ambient temperature and then held at 5-IOC. Also, immature cultures survived best at 15-20'C. Further, Sterling and Platzer reported that by adjusting the pH of water to 4.5 in R. *culicivorax* cultures, the nematophagous fungus C. *anguillulae* could be effectively controlled.

Chapman and Finney reported that mature eggs of *R. culicivorax* can be stored for 12 weeks at IO°C and 21 weeks if stored during early development at 15°C and later transferred to IO°C. They also reported that eggs of *R. culicivorax* placed in a damp fluoroform product could be shipped with only a 10 per cent loss. The technique has the advantages over the standard shipment of sand cultures of being about eight times lighter, and costing less to ship in addition to reducing shipping losses.

In an actual test of the mass propagation system, sufficient *R. culicivorax* were reared to treat 144,000 m2 of breeding area in El Salvador. The necessary inoculum required the exposure of 1.6 million first-instar *Cx. quinquefasciatus* to 137 million preparasites (I: 14 ratio) each week for 6 weeks. The system produced an average of 13.7 g (about 2000 g- 1) of postparasitic nematodes per rearing tray (20,000 mosquitoes), a total of 6392 g for the 6-week period and 435 cultures.

In Vitro **Propagation**

The main attribute of successful *in vitro* rearing systems is that they will permit the production of species -not currently available by *in vivo* methods. Also, *in vitro* systems offer the potential for reduction of costs over *in vivo* systems, and better product standardization and formulation. The development of a successful *in vitro* rearing system for mermithids has been slow because of a lack of readily available material. Consequently, most basic *in vitro* research has been done with *R. culicivorax.* Another major problem involves the nature of food ingestion in mermithids. Unlike most nematodes, mermithids lack a functional gut and must rely on transcuticular uptake of nutrients. Since these nematodes feed only during their parasitic phase while in the hemocoel of the host, nutritional products must be in a form that can be readily absorbed through a selective and delicate cuticle. Thus, they are very sensitive to mechanical damage and to hypo- and hypertonic changes.

In initial studies with *R. culicivorax, Sanders et al.* reported growth of 1-4 mm in length after 15 days and 5-7 mm after 25 days and some stichosome and trophosome development when preparasites were placed in Schneider's *Drosophila* medium with 10 per cent fetal calf serum at pH of 6.5-7.0 and 25°C. Growth rates were about one-third that of nematodes grown *in vivo.*

More recently, Castillo *et al.* tested more than 50 combinations of vertebrate and invertebrate tissue culture media and microbiological media. Slow growth and limited development of internal structures were obtained with various supplemented Grace's tissue culture media and Schneider's *Drosophila* media. Nematodes attained a stage of development after 3-4 weeks comparable to that attained in the host after 4-5 days.

The most successful attempts to rear *R. culicivorax in vitro* were reported by Finney. She reported that the best results were obtained using Grace's medium containing 10 per cent fetal calf serum at 26°C with osmotic pressure adjusted to 240 mosm and pH 6.4-6.5. Slow growth of the nematodes occurred over a 6-week period during which only juvenile females developed, but storage material in the trophosome was lacking. Finney has since reduced the time to reach parasitic maturity to 3 weeks with the trophosome filled with material. However, she was still unable to produce male nematodes.

BIO-INSECTICIDES AND PESTICIDES

Today the rapid increase in population and demand of food materials has initiated the large use of insecticides and pesticides. These toxic chemical insecticides and pesticides are resulting in harmful effects and biomagnifications due to which our environment is continuously polluting and fertile lands are acquiring infertility. No doubt they are providing hopeful results in eradication of insects, pests and diseases but they are also killing useful organism present in soil due to which the fertility of the soil is rapidly declining.

The conventional farming practices which uses chemical methods to kill both useful and harmful life forms indiscriminately, resulting in the malfunctioning of food chain and food web. Bio-control is the best method to cope with the losses done by the chemicals.

In these method insects, pests and pathogens are removed using biological methods without harming the environment and other organism. This is based on natural predation rather than introduced chemicals. The use of bio-insecticides and pesticides also comes under this category. Today due to awareness about the harmful effects of the chemical insecticides and pesticides, most of the farmers are diverting towards the organic farming. In our local area many such plants, waste matter etc. are available from which these bio-insecticides and pesticides can be prepared by using natural means only. Conventional'pesticides are generally synthetic materials that directly kill or inactivate the pest.

Being single chemical entity, chemical pesticides have resulted in increased resistance in pests. Bio pesticides also known as Biological Pesticides are pesticides derived from natural materials as animals, plants, bacteria, and certain minerals. Bio pesticides are less toxic and also reduce the pollution problems caused by conventional pesticides.

The use of bio-insecticides and bio-pesticides also fall under this category only. Organic agriculture is a unique production management system which promotes and enhances agro-ecosystem health, including biodiversity, biological cycles and soil biological activity, and this is accomplished by using on-farm agronomic, biological and mechanical methods in exclusion of all synthetic off-farm inputs.

Organic farming generally produces somewhat lower yields but sustains better yields during drought years, allowing it to reap higher yields in some cases. Studies thus far have shown that organic farming requires less water, uses few and always natural pesticides, prevents soil erosion, leaches dramatically fewer nitrates, and has been shown to have improved nutrient qualities including as much as double the flavonoids, an important antioxidant. This project is mainly based on the preparation, usage, effects and advantages of bio-insecticides and bio-pesticides.

Different types of bio-insecticides and bio-pesticide:

- Cow Urine
- Fermented curd water
- Dashparni Extract
- Neem -cow urine extract
- Mixed leaves extract
- Chili –garlic extract
- Broad spectrum formulation 1
- Broad spectrum formulation 2

Cow Urine Extract

In India, as farming goes hand in hand with cattle rearing, therefore the cow urine is available easily in the rural areas. It has acquired an important place in the hindu religion since ancient times. Any religious ceremony and rituals can't be performed without cow urine which is commonly known as gomutra. We will here use the cow urine for the control of pests and as a growth promoter for the growing crops.

Preparation and Application

Collect 5 litre of pure and concentrated cow urine and dilute it with 40 liters of water and spray it using the sprayer in one hectare of field at the evening time.

Effects and Advantages

- Due to high content of urea in it which is toxic to most of the organisms, the pests and insects etc. will not attack the leaves and buds of the crop plants.
- Due to pungent and bad smell of the extract most of the pests and insects which are attracted due to nectar and fragrance get repelled, preventing the plant.

Fermented Curd Water

In many states of rural India milk production is the main occupation associated with the farming also known as dairy management. The dairy products can also prove a major alternative in controlling the insects and pests in farmlands. One of the dairy product known as butter milk or Chaach can be used for this purpose.

It is made by the centrifugation of curd in which byproduct ghee (fats) is also obtained. There is much greater demand of butter milk in the summer season and it gets distasteful if it is kept for more than two days. This distasteful butter milk can be used to make the extract.

Preparation and Application

Take the distasteful butter milk and add equal amount of water in it. Keep it for two days in a semi-shaded place. Now take it and add 40 litre of water in 10 litre of the extract to form 50 litre of the solution. Spray this solution in 1 hectare of field in such a way that all plants get bath in the fogging spray in the early morning time.

Effects and advantages:

- The fermented butter milk contains billions of bacteria that suppress the growth of other bacteria, fungi and protozoa. The inhibitors released by these bacteria works against the unwanted pathogens.
- Moreover, fermented butter milk also contains the useful nutrients for the growth and maturation of the crop plant.

DASHPARNI EXTRACT

Neem leaves 5 kg, vitex negundo leaves 2 kg, Aristolochia leaves 2 kg, papaya (carica papaya) 2 kg, Tinospora cordifolia leaves 2 kg, Annona squamosa (Custard apple) leaves 2 kg, Pongamia pinnata (Karanja) leaves 2 kg, Ricinus communis (Castor) leaves 2 kg, Nerium indicum 2 kg, Calotropis procera leaves 2 kg, Green chili paste 2 kg, Garlic paste 250 gm, Cow dung 3 kg and Cow Urine 5 lit. Crush all of the above material and mix it in 200 liters of water to ferment for one month. Shake this mixture regularly three times a day. Filter it after one month. This extract can be stored up to 6 months.

Application: Take this extract in fogging machine and apply it on the plants. The above prepared extract is sufficient for one acre crop.

Effects and Advantages

- The extract prepared is very useful for the control of wide variety of pests such as Thrips, Leaf folder, Leafhopper etc.
- It contains neem which has ovi-position deterrent effect, cow dung and cow urine which has bio-fertilizer role also helps in enhancing useful microbial activity in soil.
- It can be prepared easily and can be stored for a large time interval for further use.

Neem-cow Urine Extract

Materials Required: 5 kg of neem leaves, 5 litre of cow urine, 2 kg of cow dung, 100 litre of water

Method of Preparation

Crush the above substances and ferment for 24 hours with intermittent stirring, filter and squeeze the extract and dilute to 100 liters of water. Use this extract to fill in the spray machine and spray it over one acre of the crop.

Effects and Advantages

- Neem is a broad spectrum pesticide and even compatible with most chemical pesticides. Neem works by intervening at several stages of the life of an insect. It may not kill the pests instantaneously but incapacitate it in several other ways.
- Neem acts in various ways like as anti-feedant, as repellent, as growth inhibitor and also has ovi-position deterrent effect.
- Cow urine has high content of urea in it which is toxic to most of the organisms, the pests and insects etc.

Mixed Leaves Extract

3 kg of neem leaves, 10 liters of cow-urine, custard apple leaves 2 kg, 2 kg papaya leaves, 2 kg pomegranate leaves, 2 kg guava leaves.

Crush all of the above recommended material and add 5 liters of water in it. Boil the above mixture 5 times after some intervals of time till the mixture becomes half of the initial. Keep it for 24 hours, filter and squeeze the extract. This can be stored in bottles for 6 months. Dilute the above prepared extract 2-2.5 liters in 50 liters of water. Fill the diluted solution in fogging machine and spray it in one acre area.

Effects and Advantages

- Guava leaves has achieved a special place in ayurveda due to its medical use for curation of some diseases.
- Pomegranate leaves have resistive nature against many insects and pests due to presence of some special compounds in it.

Chili-garlic Extract

Ipomea leaves 1 kg, 500 gm hot chili, 500 gm garlic, 5 kg neem leaves, 10 litre cow urine Crush the above recommended material and boil the suspension till it becomes half of the initial. Filter and squeeze the above extract and store it in glass or plastic bottles. Take 2-3 liters of the extract and dilute it with 50 liters of water. Now mix it thoroughly and use it as a foliar spray for one acre of crop.

Effects and advantages:

- Garlic contains sulphur which is an antibacterial.
- Chili has the property to avoid fungal and bacterial infection due to its preservative property.

Broad Spectrum Formulation 1

3 kg fresh crushed neem leaves, 1 kg neem seed kernel powder, 10 lit of cow urine, 500 gm green chilies and 250gm of garlic

Take a copper container and mix 3 kg neem leaves, neem seed kernel powder with 10 liters of cow urine. Seal the container and allow the suspension to be ferment for 10 days. After 10 days boil the suspension, till the volume becomes half of the initial.

Now ground 500 gm of green chilies in 1 litre of water and keep overnight. In another container crush 250 gm of garlic, add 1 litre of water and keep it overnight. Next day mix the above boiled extract, chili extract and garlic extract. Mix it vigourously and filter it. Store the concentrate in glass or plastic containers. Dilute 250 ml of concentrate with 10 liters of water and mix it thoroughly. Now use it as a foliar spray on the crop.

Effects and Advantages

- Due to the formation of complex compounds of Cu in this formulation which are poisonous for many bacteria, fungi, protozoa etc., it is very useful for microbial control.
- The compounds produced by the fermentation of the above extract are poisonous to many micro-organisms and insects.

Broad Spectrum Formulation 2

5 kg neem seed kernel powder, 1kg Karanj seed powder, 5 kg chopped leaves of besharam (Ipomeas sp.), 5kg chopped neem leaves, 10-12 lit of cow urine Suspend 5 kg neem seed kernel powder, 1kg Karanj seed powder, 5 kg chopped leaves of besharam (Ipomeas sp.) and 5kg chopped neem leaves in a 20 litre drum. Add 10-12 lit of cow urine and fill the drum with water to make 150 lit. Seal the drum and allow it to ferment for 8-10 days. After 8 days mix the contents and distil in a distiller.

Distillate obtained from 150 litre liquid will be sufficient for one acre. Dilute in appropriate proportion and use as foliar spray.

Effects and Advantages

- Neem effectively controls common pests like Thrips, Whitefly, Leaf folder, Bollworms, Aphids, Jassids, Pod borer, Fruit borer, Stem borer, Leafhopper, Caterpillars, Diamond back moth.
- Karanj contains poisonous chemicals in it and can be used to check microbial growth.

PLANT PATHOGENS AND PARASITIC FUNGI

In several countries people eat plant material infected with plant pathogenic fungi. Maize cobs infected with the smut fungus *Ustilago maydis* are consumed in large quantities in Mexico, both fresh and canned. They are known locally as *huitlacoche* or *cuitlacoche. U. maydis* is a microfungus: it does not form a visible fruiting body and the only signs of its presence are a mass of dark spores. The cobs appear to become sweeter as the result of fungus attack, and similar changes have been noted for the edible rust fungus *Cronartium conigenum* on pines in Mexico.

Other examples include: *Ustilago esculenta* on wild rice; *Sporisorium cruenta* on sorghum in China, personal communication: *Eating* Sporisorium cruenta *in China*); winged bean infected by *Synchytrium psophocarpi* in Indonesia.

Hypomyces lactifluorum is a parasite macrofungus that grows on other macrofungi. It is eaten from Canada through to Guatemala and completes the range of ecological niches occupied by wild edible fungi.

LOCAL AND SCIENTIFIC NAMES

Local names have been well documented in Mexico, China and can be checked online for Malawi 4 against the equivalent scientific names. Each of these countries has a rich lexicon of names and terms, a sign of the importance of wild edible fungi to rural people. Some local names have been adopted more widely, particularly for valuable edible fungi. *Boletus edulis* is commonly referred to by its French (*cèpe*) or Italian name (*porcino*– plural *porcini*), and *Tricholoma matsutake* by its Japanese name of *matsutake*. The system of scientific names aims to remove doubt about the fungus being described. A person with

Cantharellus cibarius in Nepal knows they have the same fungus as someone in Mozambique, assuming both have been accurately identified. The scientific name or binomial has two parts. The first name is the genus (*Cantharellus*) followed by the species name (*cibarius*). Named varieties exist for some species but their scientific validity is often uncertain.

Local names for edible fungi are based on shape, taste and other properties that are distinctive or important to people. The lichen (*Umbilicaria esculenta*) and an edible fungus (*Auricularia auricula-judae*) have similar common names in Hunan – *Yan-er* (ear of a rock) and *Mu-er* (ear of wood) respectively. This identifies where they grow and can be collected. Mycologists are sometimes wary of local classifications because they are based on scientifically unreliable characters.

Local names provide important clues to the uses and importance of edible fungi to people and there is much to be gained from their study. Local names allow researchers to learn about collecting practices, to analyse markets and to talk with forest managers and others who lack formal training in science and are unfamiliar with genera and species names. Examples of ethnographic studies involving wild useful fungi are list. Guidelines for conducting such studies are available from a number of different sources.

Local and scientific classifi-cations serve two different groups of people and neither is infallible. Edible species of *Boletus* are not eaten in parts of the United Republic of Tanzania, for example, reflecting local custom rather than scientific fact. Field guides often disagree on which species are edible, either because they are cautious about recommending species that require pre-cooking or because the authors are unaware of local customs in different parts of the world.

What is clear, however, is that there are many poorly described species sold and collected for personal use in developing countries. The rate of discovery is directly related to funding for projects and the ability to draw upon mycological expertise from different countries. Work in the United Republic of Tanzania, Mozambique and Malawi, Burundi and Benin emphasizes the richness of the tropical, edible mycota and how much remains to be done. In the absence of such mycological expertise local names can provide useful information, particularly if dried specimens are available for later examination.

An accurately identified specimen with a scientific name for that species ensures that any new knowledge can be reliably used. A scientific name is the most useful way of finding out whether a species is edible or poisonous, or if it has medicinal or other useful properties. An importer does not need to know if the *pied de mouton* from Bulgaria is *Hydnum repandum* since the genus contains only edible species, but an Italian buyer will pay less for the ordinary *Tuber sinosum* from China compared with other more valuable species. In this instance a scientific name reliably and uniquely describes the fungus in question, for which information can be gleaned from the literature. The scientific names for fungi are constantly changing – an indication of how much there is still to discover about the diversity of species. New names are proposed and generic boundaries adjusted, both as the result of new discoveries and a revision of the relationships between species. When a new species is proposed it is judged against guidelines and rules drawn up and regularly revised by scientists. The correct publication of a new name does not mean that scientists agree on its taxonomic status. The boundaries between genera and species are open to different interpretations and that is why there are "preferred" rather than "correct" scientific names for fungi.

These changes and uncertainties have important practical consequences for people using wild edible fungi. People have to be aware that a species was previously known by a different name or synonym when searching for information: *Termitomyces albuminosus* was once known as *Collybia albuminosa.* Other changes are less dramatic. *Lentinus edodes,* or *shi'itake* now has the preferred name of *Lentinula edodes.* The older "non-preferred" name is still regularly used in publications. Opinions are still divided as to whether *Coriolus* species with medicinal properties should be renamed *Trametes. Auricularia auricula-judae,* the "preferred name", appears variously as *Hirneola auricula-judae* and *Auricularia auricula.*

The preferred names of wild edible fungi that are still commonly referred to by other names. Common spelling mistakes also appear in publications; even minor differences can cast doubt on the identity of a fungus.

The *Dictionary of the fungi* is a standard reference that is regularly revised to list all genera of fungi. Index Fungorum, an Internet resource, allows users to check the preferred or non-preferred status for species names and to find synonyms. This is of considerable practical benefit, although Index Fungorum lacks the backing required to answer fully queries about which scientific name to use for wild edible fungi. This practical need has still to be addressed by the scientific community.

AS PATHOGENS AND PARASITES

Many fungi are parasites on plants, animals (including humans), and other fungi. Serious pathogens of many cultivated plants causing extensive damage and losses to agriculture and forestry include the rice blast fungus *Magnaporthe oryzae,* tree pathogens such as *Ophiostoma ulmi* and *Ophiostoma novo-ulmi* causing Dutch elm disease, and *Cryphonectria parasitica* responsible for chestnut blight, and plant pathogens in the genera *Fusarium, Ustilago, Alternaria,* and *Cochliobolus.* Some carnivorous fungi, like *Paecilomyces lilacinus,* are predators of nematodes, which they capture using an array of specialized structures such as constricting rings or adhesive nets.

Some fungi can cause serious diseases in humans, several of which may be fatal if untreated. These include aspergilloses, candidoses, coccidioidomycosis, cryptococcosis, histoplasmosis, mycetomas, and paracoccidioidomycosis. Furthermore, persons with immuno-deficiencies are particularly susceptible to disease by genera such as *Aspergillus, Candida, Cryptoccocus, Histoplasma,* and *Pneumocystis.* Other fungi can attack eyes, nails, hair, and especially skin, the so-called dermatophytic and keratinophilic fungi, and cause local infections such as ringworm and athlete's foot. Fungal spores are also a cause of allergies, and fungi from different taxonomic groups can evoke allergic reactions.

HUMAN USE

The human use of fungi for food preparation or preservation and other purposes is extensive and has a long history. Mushroom farming and mushroom gathering are large industries in many countries. The study of the historical uses and sociological impact of fungi is known as ethnomycology. Because of the capacity of this group to produce an enormous range of natural products with antimicrobial or other biological activities, many species have long been used or are being developed for industrial production of antibiotics, vitamins, and anti-cancer and cholesterol-lowering drugs. More recently, methods have been developed for genetic engineering of fungi, enabling metabolic engineering of fungal species. For example, genetic modification of yeast species —which are easy to grow at fast rates in large fermentation vessels—has opened up ways of pharmaceutical production that are potentially more efficient than production by the original source organisms.

Antibiotics

Many species produce metabolites that are major sources of pharmacologically active drugs. Particularly important are the antibiotics, including the penicillins, a structurally related group of â-lactam antibiotics that are synthesized from small peptides. Although naturally occurring penicillins such as penicillin G (produced by *Penicillium chrysogenum*) have a relatively narrow spectrum of biological activity, a wide range of other penicillins can be produced by chemical modification of the natural penicillins. Modern penicillins are semisynthetic compounds, obtained initially from fermentation cultures, but then structurally altered for specific desirable properties.

Other antibiotics produced by fungi include: griseofulvin from *Penicillium griseofulvin* used to treat dermatophyte infections of the skin, hair and nails; cyclosporins, commonly used as an immunosuppressant during transplant surgery; and fusidic acid, used to help control infection from methicillin-resistant *Staphylococcus aureus* bacteria. Widespread use of these antibiotics for the treatment of bacterial diseases, such as tuberculosis, syphilis, leprosy, and many others began in the early 20th century and continues to play a major part in anti-bacterial chemotherapy. In nature, antibiotics of fungal or bacterial origin appear to play a dual role: at high concentrations they act as chemical defense against competition with other microorganisms in species-rich environments, such as the rhizosphere, and at low concentrations as quorum-sensing molecules for intra-or interspecies signaling.

Cultured Foods

Baker's yeast or *Saccharomyces cerevisiae,* a single-celled fungus, is used to make bread and other wheat-based products, such as pizza dough and dumplings. Yeast species of the genus *Saccharomyces* are also used to produce alcoholic beverages through fermentation. Shoyu koji mold (*Aspergillus oryzae*) is an essential ingredient in brewing Shoyu (soy sauce) and sake, and the preparation of miso, while *Rhizopus* species are used for making tempeh. Several of these fungi are domesticated species that were bred or selected based on their capacity to ferment food without producing harmful mycotoxins, which are produced by very closely related *Aspergilli.* Quorn, a meat substitute, is made from *Fusarium venenatum.*

Medicinal Use

Certain mushrooms enjoy usage as therapeutics in traditional and folk medicines, such as Traditional Chinese medicine. Notable species with a well-documented history of use include *Agaricus blazei, Ganoderma lucidum,* and *Cordyceps sinensis.* Research has identified compounds produced by these and other fungi that have inhibitory biological effects against viruses and cancer cells. Specific metabolites with biological or antimicrobial activities, such as polysaccharide-K, ergotamine, and â-lactam antibiotics, are routinely used in clinical medicine. The shiitake mushroom is a source of lentinan, a clinical drug approved for use in cancer treatments in several countries, including Japan. In Europe and Japan, polysaccharide-K (brand name Krestin), a chemical derived from *Trametes versicolor,* is an approved adjuvant for cancer therapy.

Edible and Poisonous Species

Edible mushrooms are well-known examples of fungi. Many are commercially raised, but others must be harvested from the wild. *Agaricus bisporus,* sold as button mushrooms when small or Portobello mushrooms when larger, is a commonly-eaten species, used in salads, soups, and many other dishes. Many Asian fungi are commercially grown and have increased in popularity in the West. They are often available fresh in grocery stores and markets, including straw mushrooms (*Volvariella volvacea*), oyster mushrooms (*Pleurotus ostreatus*), shiitakes (*Lentinula edodes*), and enokitake (*Flammulina* spp.).

There are many more mushroom species that are harvested from the wild for personal consumption or commercial sale. Milk mushrooms, morels, chanterelles, truffles, black trumpets, and *porcini* mushrooms (*Boletus edulis*) (also known as king boletes) demand a high price on the market. They are often used in gourmet dishes.

Certain types of cheeses require inoculation of milk curds with fungal species that impart a unique flavour and texture to the cheese. Examples include the blue colour in cheeses such as Stilton or Roquefort, which are made by inoculation with *Penicillium roqueforti*. Molds used in cheese production are non-toxic and are thus safe for human consumption; however, mycotoxins (*e.g.*, aflatoxins, roquefortine C, patulin, or others) may accumulate due to growth of other fungi during cheese ripening or storage.

Many mushroom species are poisonous to humans, with toxicities ranging from slight digestive problems or allergic reactions as well as hallucinations to severe organ failures and death. The most deadly mushrooms belong to the genera *Inocybe*, *Cortinarius*, and most infamously, *Amanita*. The latter genus includes the destroying angel *(A. virosa)* and the death cap, the most common cause of deadly mushroom poisoning. The false morel is occasionally considered a delicacy when cooked, yet can be highly toxic when eaten raw. *Tricholoma equestre* was considered edible until being implicated in serious poisonings causing rhabdomyolysis. Fly agaric mushrooms also cause occasional non-fatal poisonings, mostly as a result of ingestion for use as a recreational drug for its hallucinogenic properties. Historically, fly agaric was used by different peoples in Europe and Asia and its present usage for religious or shamanic purposes is reported from some ethnic groups such as the Koryak people of northeastern Siberia.

As it is difficult to accurately identify a safe mushroom without proper training and knowledge, it is often advised to assume that a wild mushroom is poisonous and not to consume it.

Pest Control

In agriculture, fungi may be useful if they actively compete for nutrients and space with pathogenic microorganisms such as bacteria or other fungi via the competitive exclusion principle, or if they are parasites of these pathogens.

For example, certain species may be used to eliminate or suppress the growth of harmful plant pathogens, such as insects, mites, weeds, nematodes and other fungi that cause diseases of important crop plants. This has generated strong interest in practical applications that use these fungi in the biological control of these agricultural pests. Entomopathogenic fungi can be used as biopesticides, as they actively kill insects.

Examples that have been used as biological insecticides are *Beauveria bassiana*, *Metarhizium anisopliae*, *Hirsutella* spp, *Paecilomyces* spp, and *Verticillium lecanii*. Endophytic fungi of grasses of the genus *Neotyphodium*, such as *N. coenophialum*, produce alkaloids that are toxic to a range of invertebrate and vertebrate herbivores. These alkaloids protect grass plants from herbivory, but several endophyte alkaloids can poison grazing animals, such as cattle and sheep. Infecting cultivars of pasture or forage grasses with *Neotyphodium* endophytes is one approach being used in grass breeding programmes; the fungal strains are selected for producing only alkaloids that increase resistance to herbivores such as insects, while being non-toxic to livestock.

Bioremediation

Certain fungi, in particular "white rot" fungi, can degrade insecticides, herbicides, pentachlorophenol, creosote, coal tars, and heavy fuels and turn them into carbon dioxide, water, and basic elements. Fungi have been shown to biomineralize uranium oxides, suggesting they may have application in the bioremediation of radioactively polluted sites.

Model Organisms

Several pivotal discoveries in biology were made by researchers using fungi as model organisms, that is, fungi that grow and sexually reproduce rapidly in the laboratory. For example, the one gene-one enzyme hypothesis was formulated by scientists who used the bread mold *Neurospora crassa* to test their biochemical theories. Other important model fungi are *Aspergillus nidulans* and the yeasts, *Saccaromyces cerevisiae* and *Schizosaccharomyces pombe,* each of which has a long history of use to investigate issues in eukaryotic cell biology and genetics, such as cell cycle regulation, chromatin structure, and gene regulation. Other fungal models have more recently emerged that each address specific biological questions relevant to medicine, plant pathology, and industrial uses; examples include *Candida albicans*, a dimorphic, opportunistic human pathogen, *Magnaporthe grisea*, a plant pathogen, and *Pichia pastoris*, a yeast widely used for eukaryotic protein expression.

Others

Fungi are used extensively to produce industrial chemicals like citric, gluconic, lactic, and malic acids, antibiotics, and even to make stonewashed jeans. Fungi are also sources of industrial enzymes, such as lipases used in biological detergents, amylases, cellulases, invertases, proteases and xylanases. Several species, most notably *Psilocybin mushrooms* (colloquially known as *magic mushrooms*), are ingested for their psychedelic properties, both recreationally and religiously.

Mycotoxins

Many fungi produce biologically active compounds, several of which are toxic to animals or plants and are therefore called mycotoxins. Of particular relevance to humans are mycotoxins produced by molds causing food spoilage, and poisonous mushrooms. Particularly infamous are the lethal amatoxins in some *Amanita* mushrooms, and ergot alkaloids, which have a long history of causing serious epidemics of ergotism (St Anthony's Fire) in people consuming rye or related cereals contaminated with sclerotia of the ergot fungus, *Claviceps purpurea*. Other notable mycotoxins include the aflatoxins, which are insidious liver toxins and highly carcinogenic metabolites produced by certain *Aspergillus* species often growing in or on grains and nuts consumed by humans, ochratoxins, patulin, and trichothecenes and fumonisins, which have significant impact on human food supplies or animal livestock.

Mycotoxins are secondary metabolites (or natural products), and research has established the existence of biochemical pathways solely for the purpose of producing mycotoxins and other natural products in fungi. Mycotoxins may provide fitness benefits in terms of physiological adaptation, competition with other microbes and fungi, and protection from consumption (fungivory).

Mycology

Mycology is the branch of biology concerned with the systematic study of fungi, including their genetic and biochemical properties, their taxonomy, and their use to humans as a source of medicine, food, and psychotropic substances consumed for religious purposes, as well as their dangers, such as poisoning or infection. The field of phytopathology, the study of plant diseases, is closely related because most plant pathogens are fungi.

Use of fungi by humans dates back to prehistory; Ötzi the Iceman, a well-preserved mummy of a 5,300 year old Neolithic man found frozen in the Austrian Alps, carried two species of polypore mushrooms that may have been used as tinder (*Fomes fomentarius*), or for medicinal purposes (*Piptoporus betulinus*).

Ancient peoples have used fungi as food sources – often unknowingly – for millennia, in the preparation of leavened bread and fermented juices. Some of the oldest written records contain references to the destruction of crops that were probably caused by pathogenic fungi.

History

Mycology is a relatively new science that became systematic after the development of the microscope in the 16th century. Although fungal spores were first observed by Giambattista della Porta in 1588, the seminal work in the development of mycology is considered to be the publication of Pier Antonio Micheli's 1729 work *Nova plantarum genera*. Micheli not only observed spores, but showed that under the proper conditions, they could be induced into growing into the same species of fungi from which they originated.

Extending the use of the binomial system of nomenclature introduced by Carl Linnaeus in his *Species plantarum* (1753), the Dutch Christian Hendrik Persoon (1761–1836) established the first classification of mushrooms with such skill so as to be considered a founder of modern mycology. Later, Elias Magnus Fries (1794–1878) further elaborated the classification of fungi, using spore colour and various microscopic characteristics, methods still used by taxonomists today.

Other notable early contributors to mycology in the 17th–19th and early 20th centuries include Miles Joseph Berkeley, August Carl Joseph Corda, Anton de Bary, the brothers Louis René and Charles Tulasne, Arthur H. R. Buller, Curtis G. Lloyd, and Pier Andrea Saccardo. The twentieth century has seen a modernization of mycology that has come from advances in biochemistry, genetics, molecular biology, and biotechnology. The use of DNA sequencing technologies and phylogenetic analysis has provided new insights into fungal relationships and biodiversity, and has challenged traditional morphology-based groupings in fungal taxonomy.

NEMATODE CONTROL OF FUNGI

Fungi utilise many different sources of carbon, some living and some dead. The observation that nematodes can be colonised by fungi, led to the possibility that the interaction might be used to control pest nematodes of plants and animals. Closer examination of the interaction has led to the belief that only partial control of nematodes is possible, and that in a practical sense, populations of pests may be suppressed not eliminated. In the case of plant and animal parasitic nematodes, fungi may be used to regulate nematode populations in conjunction with other processes.

PEST NEMATODES

Nematodes found in plants, animals and soil use various sources of organic energy. Plant parasites feed from plant cells, either by selecting and then establishing a single feeding site. This approach is usually known as sedentary feeding. Other plant-parasitic nematodes are migratory, moving from site to site on the root, rarely using a single cell for a long time. Migratory plant parasites can be further categorised as internal and external feeders, meaning that they migrate within or outside the root.

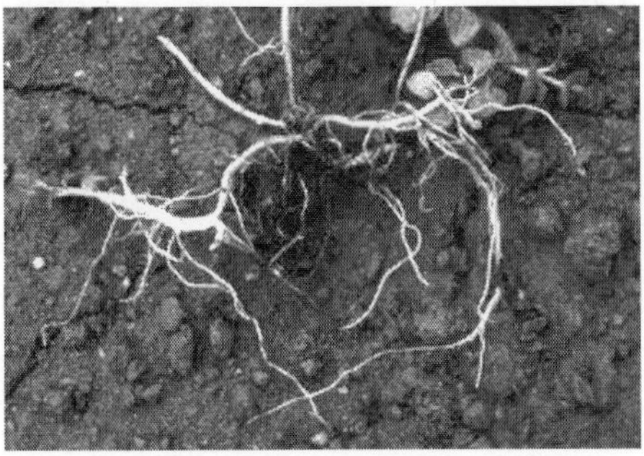

Fig. Nematode in roots

Animal parasites of agriculturally important ruminants tend to have a life cycle whereby colonisation of the host is followed by a soil phase. Larvae are consumed by grazing animals along with grass. The larvae establish a feeding location within the digestive system of the host. Adults lay eggs within the intestines, and the eggs fall to the ground with the faeces. Eggs subsequently hatch in the faeces and larvae move to blades of grass, where they are consumed.

Fungi that colonise nematodes fall into several different categories, each of which requires a different approach if it is to be used successfully to control pest nematodes. The fungi range from highly specialised nematode parasites, to generalists with no specific host. Some fungi may colonise organic matter and incidentally also colonise eggs, larvae or adult nematodes. Other fungi may trap larvae or adults and colonise their body, using the body as the primary source of nutritients.

Plant Parasitic Nematodes

Endoparasitic fungi are generally highly specific to single species or genera of nematodes, and are difficult to culture in the absence of their host. Hirsutella, Meria, Nematophthora and Nematoctonus have all been suggested as ideal biocontrol agents because of the specificity for economically important nematodes.

However, the difficulty of culturing them and lack of experimental support for their activity has limited any commercial development.

Nematode trapping fungi utilise either sticky pads or constricting rings to immobilise nematodes. The fungus then digests the internal organs of the nematode and form reproductive cells. The main fungi in this group are Arthrobotrys and Duddingtonia. These fungi are found in soil where they live on readily-available organic carbon, in the absence of nematodes. They are generalist feeders, being able to trap and digest many nematodes, not just those that infest plants. Thus, the fungi are rare, unable to exist at sufficiently high densities to control nematode populations, and parasitic on beneficial as well as pest species of nematode. Because of these difficulties, exploration of the use of trapping fungi in the field has largely ceased.

Parasitic fungi have been isolated from either eggs, larvae or adults nematodes. Several of these have been found to preferentially parasitise the nematode and thereby reduce the size of the nematode population. Research efforts in some crops are now concentrating on using fungi found locally to control local pests.

The general approach is to go to locations where nematodes have reached high densities. Parasitised individuals are extracted from soil, the fungi cultured and then tested as parasites of the pest nematode. The theoretical basis for this approach is that high densities of nematode will enable fungal parasites to increase population size dramatically. Thus by selecting fungi adapted to the host (nematode) and environment, effective and usually inundative inoculation will enable the control of the pest nematode. Biocontrol in problem fields has been achieved in a surprising number of cases. The species used include Paecilomyces lilacinus, a fungus that appears to have a wide range of potential hosts, both insect and nematode, and yet isolates have a degree of specificity for host.

Animal Parasitic Nematodes

Sheep and cattle have traditionally been drenched to control parasitic nematodes. However, nematodes are now developing resistance to the common active ingredients in drenches. Drenches using different active compounds are not being developed because the economical returns are low for the investment. Thus alternative approaches are being sought.

In one approach, nematode trapping fungi are placed into nutrient-rich faeces. As nematodes hatch, they become colonised by the fungus. The fungus Duddingtonia has been inoculated to the dung via a drench to the animal. Sclerotia of the fungus pass through the intestine, hatch in moist dung, and effectively reduce larval densities in some cases.

While further research is needed before commercial adaptation, this is one example where understanding of the biology of fungi, lateral thinking and some judicious research, has led to the development of a new approach to a serious problem.

Biocontrol of nematodes by fungi appeared to be possible when nematode trapping fungi were first discovered. However, it has taken some considerable time before their potential could be realised. Successful control relies on having sufficiently high densities of fungus where high densities of the pest are maintained. The requirement for high densities of agent to control the pest applies across all biocontrol efforts.

7

Soil Microorganisms in Biodegradation of Pesticides

INTRODUCTION

Pesticides are the chemical substances that kill pests and herbicides are the chemicals that kill weeds. In the context of soil, pests are fungi, bacteria insects, worms, and nematodes etc. that cause damage to field crops. Thus, in broad sense pesticides are insecticides, fungicides, bactericides, herbicides and nematicides that are used to control or inhibit plant diseases and insect pests. Although wide-scale application of pesticides and herbicides is an essential part of augmenting crop yields; excessive use of these chemicals leads to the microbial imbalance, environmental pollution and health hazards. An ideal pesticide should have the ability to destroy target pest quickly and should be able to degrade non-toxic substances as quickly as possible.

The ultimate "sink" of the pesticides applied in agriculture and public health care is soil. Soil being the storehouse of multitudes of microbes, in quantity and quality, receives the chemicals in various forms and acts as a scavenger of harmful substances. The efficiency and the competence to handle the chemicals vary with the soil and its physical, chemical and biological characteristics.

EFFECTS OF PESTICIDES

Pesticides reaching the soil in significant quantities have direct effect on soil microbiological aspects, which in turn influence plant growth. Some of the most important effects caused by pesticides are:

- Alterations hi ecological balance of the soil microflora,
- Continued application of large quantities of pesticides may cause ever lasting changes in the soil microflora,
- Adverse effect on soil fertility and crop productivity,
- Inhibition of N2 fixing soil microorganisms such as *Rhizobium*, *Azotobacter*, *Azospirillum* etc. and cellulolytic and phosphate solubilizing microorganisms,

- Suppression of nitrifying bacteria, *Nitrosomonas* and *Nitrobacter* by soil fumigants ethylene bromide, Telone, and vapam have also been reported,
- Alterations in nitrogen balance of the soil,
- Interference with ammonification in soil,
- Adverse effect on mycorrhizal symbioses in plants and nodulation in legumes, and
- Alterations in the rhizosphere microflora, both quantitatively and qualitatively.

PERSISTENCE OF PESTICIDES IN SOIL

How long an insecticide, fungicide, or herbicide persists in soil is of great importance in relation to pest management and environmental pollution. Persistence of pesticides in soil for longer period is undesirable because of the reasons:

- Accumulation of the chemicals in soil to highly toxic levels,
- May be assimilated by the plants and get accumulated in edible plant products,
- Accumulation in the edible portions of the root crops,
- To be get eroded with soil particles and may enter into the water streams, and finally leading to the soil, water and air pollutions.

The effective persistence of pesticides in soil varies from a week to several years depending upon structure and properties of the constituents in the pesticide and availability of moisture in soil. For instance, the highly toxic phosphates do not persist for more than three months while chlorinated hydrocarbon insecticides (eg. DDT, aldrin, chlordane etc) are known to persist at least for 4-5 years and some times more than 15 years.

From the agricultural point of view, longer persistence of pesticides leading to accumulation of residues in soil may result into the increased absorption of such toxic chemicals by plants to the level at which the consumption of plant products may prove deleterious/ hazardous to human beings as well as livestock's.

There is a chronic problem of agricultural chemicals, having entered in food chain at highly inadmissible levels in India, Pakistan, Bangladesh and several other developing countries in the world. For example, intensive use of DDT to control insect pests and mercurial fungicides to control diseases in agriculture had been known to persist for longer period and thereby got accumulated in the food chain leading to food contamination and health hazards. Therefore, DDT and mercurial fungicides has been, banned to use in agriculture as well as in public health department.

BIODEGRADATION OF PESTICIDES IN SOIL

Pesticides reaching to the soil are acted upon by several physical, chemical, and biological forces. However, physical and chemical forces are acting upon/degrading the pesticides to some extent, microorganism's plays major role in the degradation of pesticides. Many soil microorganisms have the ability to act upon pesticides and convert them into simpler non-toxic compounds.

This process of degradation of pesticides and conversion into non-toxic compounds by microorganisms is known as "biodegradation". Not all pesticides reaching to the soil are biodegradable and such chemicals that show complete resistance to biodegradation are called "recalcitrant". The chemical reactions leading to biodegradation of pesticides fall into several broad categories which are discussed in brief in the following paragraphs.

- *Detoxification:* Conversion of the pesticide molecule to a non-toxic compound. Detoxification is not synonymous with degradation. Since a single chance in the side chain of a complex molecule may render the chemical non-toxic.
- *Degradation:* The breaking down/ transformation of a complex substrate into simpler products leading finally to mineralization. Degradation is often considered to be synonymous with mineralization, *e.g.* Thirum (fungicide) is degraded by a strain of*Pseudomonas* and the degradation products are dimethlamine, proteins, sulpholipaids, etc.
- *Conjugation (complex formation or addition reaction):* In which an organism make the substrate more complex or combines the pesticide with cell metabolites. Conjugation or the formation of addition product is accomplished by those organisms catalyzing the reaction of addition of an amino acid, organic acid or methyl crown to the substrate, for *e.g.*, in the microbial metabolism of sodium dimethly dithiocarbamate, the organism combines the fungicide with an amino acid molecule normally present in the cell and thereby inactivate the pesticides/ chemical.
- *Activation:* It is the conversion of non-toxic substrate into a toxic molecule, for eg. Herbicide, 4-butyric acid (2, 4-D B) and the insecticide Phorate are transformed and activated microbiologically in soil to give metabolites that are toxic to weeds and insects.
- *Changing the spectrum of toxicity:* Some fungicides/pesticides are designed to control one particular group of organisms/ pests, but they are metabolized to yield products inhibitory to entirely dissimilar groups of organisms, for *e.g.* the fungicide PCNB fungicide is converted in soil to chlorinated benzoic acids that kill plants.

Biodegradation of pesticides/ herbicides is greatly influenced by the soil factors like moisture, temperature, PH and organic matter content, in addition to microbial population and pesticide solubility. Optimum temperature, moisture and organic matter in soil provide congenial environment for the break down or retention of any pesticide added in the soil. Most of the organic pesticides degrade within a short period (3-6 months) under tropical conditions.

Metabolic activities of bacteria, fungi and actinomycetes have the significant role in the degradation of pesticides.

CRITERIA FOR BIOREMEDIATION/ BIODEGRADATION

For successful biodegradation of pesticide in soil, following aspects must be taken into consideration.

- Organisms must have necessary catabolic activity required for degradation of contaminant at fast rate to bring down the concentration of contaminant,
- The target contaminant must be bioavailability,
- Soil conditions must be congenial for microbial/plant growth and enzymatic activity and
- Cost of bioremediation must be less than other technologies of removal of contaminants.

According to Gales (1952) principal of microbial infallibility, for every naturally occurring organic compound there is a microbe/ enzyme system capable its degradation.

STRATEGIES FOR BIOREMEDIATION

For the successful biodegradation/ bioremediation of a given contaminant following strategies are needed.

- *Passive/ intrinsic Bioremediation:* It is the natural bioremediation of contaminant by tile indigenous microorganisms and the rate of degradation is very slow.
- *Biostimulation:* Practice of addition of nitrogen and phosphorus to stimulate indigenous microorganisms in soil.
- *Bioventing:* Process/way of Biostimulation by which gases stimulants like oxygen and methane are added or forced into soil to stimulate microbial activity.
- *Bioaugmentation:* It is the inoculation/introduction of microorganisms in the contaminated site/soil to facilitate biodegradation.
- *Composting:* Piles of contaminated soils are constructed and treated with aerobic thermophilic microorganisms to degrade contaminants. Periodic physical mixing and moistening of piles are done to promote microbial activity.
- *Phytoremediation:* Can be achieved directly by planting plants which hyperaccumulate heavy metals or indirectly by plants stimulating microorganisms in the rhizosphere.

- *Bioremediation:*Process of detoxification of toxic/unwanted chemicals/ contaminants in the soil and other environment by using microorganisms.
- *Mineralization:* Complete conversion of an organic contaminant to its inorganic constituent by a species or group of microorganisms.

SOIL MICROBES

Soil looks lifeless but a quarter of a teaspoon of soil (1 gram) contains many microbes - 100 million bacteria, 1 million actinomycetes and 5 metres of fungal filaments.

Ways of Feeding

Bacteria and fungi have similar ways of feeding. Their main food is organic detritus in leaf litter, soil, dung and carrion. In contrast to invertebrate animals which engulf food and digest it within their bodies, microbes digest their food outside themselves by secreting digestive enzymes over their food. Under the action of the enzymes, organic matter breaks down into simpler molecules external to the microbe, which then re-absorbs the simpler products of digestion through its cell wall. Microbes have a vast number of enzymes to do this job – 50 to 60 different ones have been discovered. Few invertebrates possess this many digestive enzymes.

FUNGI

Most of the time fungi hide away in the murky depths of the soil and are not visible. But their microscopic, mouldy strands are growing through bits of organic matter and soil, through leaf litter, dead wood and dung and into carcasses of dead animals. Here, the fungi quietly go about their job of decomposing detritus and recycling the nutrients back to the soil.

It is only when these microscopic strands merge together to form the fruiting bodies of mushrooms and toadstools that we become aware of this invisible network beneath our feet. The spores borne on these fruiting bodies disperse in the wind, or in or on, the bodies of animals feeding on them e.g. springtails, flies.

Fungal fruiting bodies are of many different and beautiful forms: mushrooms, toadstools, bracket fungi, puffballs, earthstars, coral fungi, large boletes and delicate crinoline fungi. Fungal fruiting bodies can also be rather gross. The starfish fungus smells revolting so as to attract blowflies which feed on the squishy mess in the centre of the fungus and transport its spores from place to place. Other soil fungi don't produce such conspicuous fruiting bodies, but form their spores or fruiting bodies within the soil e.g. the edible and delicious truffle which has to be sniffed out with the truffle-hunter's pigs or dogs. Fungi grow by dividing and spreading out into a network of branches which grow through organic matter. Fungi can link different parts of the soil profile together. They can grow out from a piece of leaf litter, down into the soil, up into a dung pat and link all these together. It can absorb nutrients in one spot and transport them to another. This process is called translocation of nutrients and it is important in nutrient cycling. The networks of fungal filaments (mycelia) can explore their environment much more effectively than can soil bacteria which occur either as individual cells or as sedentary colonies.

The main food of fungi is organic matter in leaf litter, dung, soil and carrion. They (along with bacteria) are the main decomposer groups in soil because of their vast collective mass supported by the variety of digestive enzymes they can secrete onto organic matter to break it down.

Some fungi are edible and collecting wild fungi is a favourite pastime in Europe where, over the centuries, they have sorted out, by trial and error, those that are safe to eat - presumably the "errors" died! Australian fungi are a bit more problematical as we have not built up such a long-standing bed of knowledge. There are no simple tests for edibility. Some old wives tales are: they are harmless if you can peel them, if they fail to turn dark on soaking in cold water or if a silver spoon cooked with the fungi, doesn't turn black. I wouldn't depend on these guidelines for edibility. To put these rules of thumb on a more scientific basis, don't eat mushrooms with yellow or greenish coloured gills, or have caps with yellow centres or caps which are thimble-shaped when young, but which smell of disinfectant (iodine) when stored in a plastic bag. Even then, it is best to stick to the fungi you can buy in the supermarket - there are many varieties to choose from these days.

Some soil fungi form a positive relationship with plant roots. They live in roots, but also extend their filament networks out into the soil where they absorb nutrients and water which they bring back to the plant to use. What the fungus gets out of this relationship is food in the form of sugars. This mutually beneficial relationship (or symbiosis) between root and fungus is called a mychorrizal association

The biggest organism in the world is invisible!

Not all soil fungi are goodies. The potato famine in Ireland in the mid-1800's was caused by a soil fungus. One in every nine people perished from starvation. Also, some soil-borne decomposer fungi can grow in grain and peanuts when these are stressed by drought. The fungus contaminates the grain and nuts with a toxin (aflatoxin) which causes liver damage in animals and humans.

Poisonous fungi

Only a small proportion of Australian fungi are poisonous but since some fungi have fatal effects it's best to stick to the supermarket varieties. Recently, a prominent Australian newspaper in its lifestyle section had to quickly publish a correction when it got the captions to two fungi mixed up. The paper had incorrectly labelled the lethal fly agaric as "edible"!

Quirky bit

Some microscopic fungi eat animals. Hard to imagine but the nematode-eating fungi grow little loops which soil nematode worms swim through. The loop tightens like a lassoo around the worm and the fungus oozes digestive enzymes onto it before absorbing the products of digestion into itself. Other nematode-eating fungi produce sticky knobs which stick to and trap nematodes.

BACTERIA

Bacteria are tiny one-celled micro-organisms, 0.001 mm in diameter, which need a light microscope with very high magnification to see them. They are best seen with an electron microscope. Bacteria have a different way of growing to fungi. They multiply by cell division, one cell dividing to make 2 cells and so on. Sometimes the cells move apart and they live as separate individuals or they can stay together to form a colony (a few cubic mm in volume) as they do when cultured in the laboratory on an agar plate. Fungal cells are different. They divide and form branching networks of filaments (mycelia) in the soil and can explore the soil environment much more effectively than bacteria.

Bacteria form resistant spores when the climate turns harsh. They can survive droughts in a spore stage and when favourable conditions return, they can germinate and grow again. The main food of bacteria is organic matter in leaf litter, dung, soil and carrion. They (along with fungi) are the main decomposer groups in soil because of the huge number of digestive enzymes they can secrete onto organic matter to break it down.

Some soil bacteria can fix atmospheric nitrogen, ultimately making it available to plants. Some of these live free in the soil e.g. Azotobacter ,while others fix nitrogen while living in nodules on the roots of legumes e.g. Rhizobia. Here, these bacteria gain food in the form of sugars from the plants and the plant benefits through increased nitrogen fixed by the bacteria. This "win-win" relationship between bacteria and plant is called "symbiosis" - or a relationship where both organisms gain a mutual benefit from living together.

Diseases from soil bacteria

Not all soil bacteria are "goodies" as soil-borne diseases can attack plants and animals. Anthrax and tetanus, which infect domestic animals and humans, lurk in soil as does the bacterial blight of tomatoes.

FORMATION OF A SOIL AGGREGATE BY MICROBES

BACTERIA

There are two ways that bacteria could be involved in soil aggregation. One way is by producing organic compounds called polysaccharides. Bacterial polysaccharides are more stable than plant polysaccharides, resisting decomposition long enough to be involved in holding soil particles together in aggregates. The other way bacteria are involved in soil aggregation is by developing a small electrostatic charge that attracts the electrostatic charge on clay surfaces, bringing together small aggregates of soil.

Fungal Hyphae

Fungi grow in long, threadlike structures, called hyphae. The amount of aggregation in the soil has been found to relate to the length of fungal hyphae in the soil. Fungi help to form aggregates in the soil by enmeshing soil particles with their hyphae and forming cross-links between soil particles. Mycorrhizal fungi and fungi that colonise fresh organic matter are believed to be the most important for assisting with stabilisation of soil particles into aggregates.

DEGRADATION OF PESTICIDES

Chemicals, such as pesticides, added to the soil are broken down in a similar way to organic matter. They are oxidised by the microorganisms that produce the appropriate enzymes. Pesticides generally have very complex structures and most are converted to several intermediate organic compounds before finally being converted to carbon dioxide and water.

When these kinds of chemicals are degraded metabolically, the microorganisms involved gain energy and carbon from the oxidation of the chemical. This allows the microorganisms to grow and multiply, and the size of the microbial population may increase as more of the chemical is degraded. Sometimes, microorganisms capable of degrading a particular chemical grow and multiply until there is a higher proportion of these microorganisms in the soil.

When a community of microorganisms in the soil adapts in this way, the chemical can be broken down more rapidly than it would have previously. This can be a problem when rapid degradation of a pesticide may reduce its effectiveness against pests.

Some chemicals added to the soil are degraded by the microorganisms without any particular benefit to the organism such as nutrients or energy supply. The population of microorganisms remains constant while the concentration of the chemical in soil decreases. The different modes of interaction between synthetic chemicals and microor-ganisms are a result of the enormous diversity of organisms in soil.

Some Chemicals Easier to Degrade than Others

The ease with which a chemical is degraded depends on the complexity of its structure. More complex structures degrade more slowly either because fewer microorganisms in the soil produce enzymes capable of degrading them or because of the inaccessibility to microbes.

THE NITROGEN (N) CYCLE

The nitrogen cycle describes how the different forms of nitrogen in the air, soil, water and living organisms are interconnected. It is described as a cycle because the nitrogen is never lost completely; it just changes form and is held in different places. There are 5 major processes involved:

- Ammonification
- Nitrification
- Denitrification
- Nitrogen Fixation
- Nitrogen Immobilisation.

Ammonification

Ammonification is the conversion of organic forms of nitrogen (for example, nitrogen in proteins in dead plants and soil animals) to ammonium. Ammonium is an inorganic form of nitrogen that has the symbol NH_4^+. A wide range of soil organisms carry out the ammonification process. Many different types of bacteria and fungi are involved. The organisms get carbon and energy from the breakdown of organic matter, while nitrogen is released at the same time.

Nitrification

Nitrification is the process where ammonium is converted to nitrite (NO_2^-) and then to nitrate (NO_3^-). The organisms involved in nitrification are a relatively small number of species in soil. They are autotrophic, which means they get energy from the chemical transformation of NH_4^+ to NO_2^-, while any needed carbon is obtained from CO_2. One group of organisms converts ammonium to nitrite and another group of organisms converts nitrite to nitrate.

Denitrification

Denitrification is the process where oxides of nitrogen (*e.g.,,* nitrate and nitrite) are converted into gaseous nitrogen and are removed from the soil system. Examples are the gas nitrous oxide (NO_2), and nitrogen gas (N_2), which is the form of nitrogen most common in the atmosphere. Many species of microorganisms are capable of denitrification of soil. Denitrification occurs mainly when there is little or no oxygen in the soil, such as when the soil is waterlogged. However, if the waterlogging is only temporary, the denitrification process will stop when the soil dries.

Nitrogen Fixation

Nitrogen fixation is the conversion of nitrogen gas (N_2) to ammonium (NH_4^+)- either by free living bacteria in soil or water, or by bacteria in symbiotic association with plants (eg legume symbiosis).

Nitrogen Immobilization

This is the process whereby nitrogen is incorporated into the microbial cells and 'tied-up' in the 'microbial pool' of nitrogen. Immobilisation occurs in parallel with ammo-nification.

THE SOIL HABITAT

Organisms Located in Soil

Most organisms are found in the top layers of soil, usually the top 2-3 centimetres, since, this is typically where most of the organic matter is. Organisms do occur to depths of several kilometres below the soil surface, but the types of organisms that occur this far down are not the same as those close to the surface. The organisms in soil are often commonly found close to root surfaces in the rhizosphere, within living and dead roots, on soil particles, or amongst aggregates of soil particles.

Earthworms and other soil animals are able to move through most of the top layers of soil. Fungi can form a mat of hyphae, which can extend centimetres or even metres through the soil. They can also form a network of hyphae inside soil aggregates. Bacteria tend to accumulate inside soil aggregates because they are less likely to be eaten by soil animals such as protozoa and mites in this environment. Bacteria can be carried down further into the soil in water that is percolating downwards, but generally they do not move far.

Soils that are clayey often have more bacteria than sandy soils because the clay creates lots of small pores (spaces) which offer protection for bacteria. Sandy soils with fewer aggregates and small pores are less suitable habitats for bacteria and fungi unless a large amount of organic matter is added to the soil.

Characteristics of the Rhizosphere

The rhizosphere is the region of soil that is immediately adjacent to and affected by plant roots. The rhizosphere is a very dynamic environment where plants, soil, microorganisms, nutrients and water meet and interact. The rhizosphere differs from the bulk soil because of the activities of plant roots and their effect on soil organisms.

A major characteristic of the rhizosphere is the release of organic compounds into the soil by plant roots. These compounds, called exudates, make the environment of the rhizosphere very different from the environment in the bulk soil. The exudates increase the availability of nutrients in the rhizosphere, and also provide a carbon source for heterotrophic microorganisms. The exudates cause the number of microorganisms to be far greater in the rhizosphere than in the bulk soil. Their presence attracts larger soil organisms that feed on microorganisms and the concentration of organisms in the rhizosphere can be up to 500 times higher than in the bulk soil. However, their growth and reproduction is even higher when grazing by predators is taken into account. This grazing helps release nutrients in microbial pool back into the soil. Thus the turnover of nutrients can be more important than the mount of microbes present at any point in time.

Another characteristic of the rhizosphere is the uptake of water and nutrients by plants. Water is drawn from the surrounding soil towards the roots. The balance between the movement of water and nutrients towards the roots and their removal from the soil by roots means that their concentration in the rhizosphere is usually very different from what it is in the bulk soil. This can affect microbial growth and activity.

Rhizoplane

The rhizoplane is the surface of plant roots in the soil. The rhizoplane is the site of water and nutrient uptake and the release of exudates into the soil. Like the rhizosphere, the rhizoplane is a constantly changing environment. As roots grow they cast off dead cells and navigate around soil particles, making the rhizoplane highly irregular, blurring the dividing line between the root surface and the soil.

Organisms in the Rhizosphere Influence Plant Roots

1. Organisms in the rhizosphere can affect the plant roots by altering the movement of carbon compounds from roots to shoots. This alteration occurs when organisms compete with root cells for carbon that is fixed by photosynthesis.
2. The burrows of earthworms in soil provide an easy route for plants roots as they grow through the soil.
3. Various root microorganism associations can increase nutrient uptake by plants in nutrient poor environments, such as symbiotic (*e.g.*, mycorrhizal) and other specific (*e.g.*, nitrogen fixing) associations and rhizobia.
4. Some soil organisms are pathogenic and attack living plant roots.
5. Azospirillum is a bacterium that commonly lives in the rhizosphere of grasses. Some strains of this organism produce hormones that stimulate plant growth. Other strains of Azospirillum fix atmospheric nitrogen and may or may not make it available to the grass.

Soil Organisms Dependent on each Other

The soil food web is a way of illustrating the way in which soil organisms relate to each other based on what they eat. The soil food web begins with organic matter, such as crop residues, pasture or any animal or plant material in the soil. Bacteria and fungi consume organic matter and are, in turn, consumed by nematodes, protozoa, earthworms, collembola and some mite species.

Nematodes and protozoa are consumed by some mites. Mites and collembola are eaten by beetles and ants. It is important to remember that the soil is a very complex habitat, in which the diversity of organisms is greater than that of the most diverse plant or animal community. This also means that it s not easy or accurate to generalize about all soil food web processes because the extent to which they occur will be different in different soils, at different times. The same processes (eg. mineralization) can be carried out by different groups of microorganisms.

SOIL ANIMALS

Soil animals are an extremely diverse group of organisms. Soil animals are grouped roughly according to their size, into three groups. The first group is the microfauna. These are the smallest of the soil animals ranging from 20 - 200 μm. The main soil animals in this group are protozoa. The mesofauna is the next largest group and range in size from 200 μm - 10mm. The most important animals in this group are mites, collembola (or spring tails) and nematodes. The macrofauna contain the largest soil animals such as earthworms, beetles and termites. Generally, the most common soil animals are protozoa, nematodes, mites and collembola.

Importantance of Soil Animals

Soil animals perform several functions in soil that make them a vital part of all ecosystems, including agriculture. Soil animals are involved in:

- Degradation of organic matter and mineralisation of nutrients,
- Controlling populations of pathogens,
- Improving and maintaining soil structure and
- Mixing organic matter through the soil.

Decomposition of Organic Matter by Soil Animals

Soil animals contribute directly to nutrient cycling in soil when they release mineralised nutrients in their excreta. However, most of their contributions are indirect, by:

- Grazing on the microbial biomass, which alters the rate at which organic matter breaks down
- Fragmenting organic matter and increasing its surface area for attack by microorganisms,
- Controlling the grazing pressure of nematodes on microorganisms,
- Mixing soil and organic matter and
- Introducing microorganisms onto fresh organic matter.

Soil Structure Affected by Soil Animals

Soil animals have an important role in the formation of soil structure. Soil animals improve soil structure by forming channels and pores, concentrating fine soil particles together into aggregates and by fragmenting and mixing organic matter through soil.

How Soil Animals Survive in Soil

Soil animals are found in almost every environment on earth, including Antarctica! Different species require different conditions to grow and survive but all soil animals, require sufficient carbon and nutrients, moisture, oxygen and an optimum pH and temperature. The optimum pH and temperature levels vary between species. Some organisms do not survive dry or very cold conditions, but they may leave eggs in the soil that hatch when conditions become more favourable. Other soil animals remain in the soil in an inactive state and become active again when conditions become favourable. Larger soil animals such as earthworms may live deeper in the soil when there are unfavourable conditions near the soil surface.

SOIL FUNGI

Fungi are primarily organisms that cannot synthesise their own food and are dependent on complex organic substances for carbon. Specialized fungi can be pathogenic on the tissues of plants, while others form mutually beneficial relationships with plants and assist in direct nutrient supply to the plants (*e.g.*, mycorrhizal associations).

Many fungi play a very important role in the recycling of important chemical elements that would otherwise remain locked up in dead plants and animals. In the decomposition of plant debris, certain fungi are particularly important because of their ability to derive their carbon and energy requirements from the break down of dead and decaying plant cell walls, cellulose and lignin. They are much less dependent on water than other microorganisms, but interactions with other microbes, temperature and nutrient availability will have an effect on their activity. Fungal activity is greatest in decomposing leaves and wood, and tends to diminish in the later stages of decomposition when bacteria become more dominant.

Mycorrhizal Fungi

Mycorrhizas are associations between fungi and plant roots that can be beneficial to both the plant and the fungi. The fungi link the plant with soil by acting as agents of nutrient exchange. The fungi receive carbohydrates as energy from the host plant root whilst nutrients such as phosphorus and zinc are passed back into the plant roots from the soil.

Mycorrhizal associations may also reduce attack from root pathogens and increase the tolerance of the plant to adverse conditions such as heavy metals, drought, and salinity. In general, mycorrhizas play an important role in sustainable plant productivity and maintenance of soil structure. Mycorrhizal associations occur on almost all terrestrial plants and are not as plant-specific as other plant-microbe associations that formed between some plants (*e.g.,*, legumes) and bacteria (*e.g.*, rhizobia).

Different Types of Mycorrhizal Associations

There are four main kinds of mycorrhizal fungi: arbuscular, ectomycorrhizal, ericoid and orchid mycorrhiza. Most agricultural plants, vegetables and orchard plants form arbuscular mycorrhizal associations. Ectomycorrhizal associations are less common in disturbed ecosystems and are more common on perennial plants than annuals. Ornamental plants form associations from each of the four groups and orchid mycorrhizal associations are formed only by orchids.

Vesicular Arbuscular Micorrhizae

The most common type of mycorrhizas are the arbuscular mycorrhizas. They are found in natural ecosystems as well as in agricultural areas, are common on both perennials and annuals, and form associations with most agricultural plants. Two exceptions among agricultural crops are canola (oilseed rape) and lupin.

Arbuscules are believed to be the major site where the carbon and nutrient exchange between plant and fungus occurs. Named because of their "tree-like" structure, arbuscules are created by repeated branching of hyphae once they enter a cell within a plant root.

Vesicles are structures formed inside a cell within the plant's roots. They can be regular or irregular in shape and many times wider than the hyphae on which they form. Therefore, vesicles are usually very distinctive and in some species can resemble spores. These structures are known as a place for storing nutrients. Only three of the five genera of arbuscular mycorrhizas (Glomus, Acaulospora and Entrophosphora) form vesicles.

The fungi are grouped according to the size, shape and wall structure of their spores. The spores are approximately spherical or ovoid in shape and usually have thick walls which allow them to survive harsh environmental conditions. When soil conditions are favourable, the spores germinate and hyphae grow from the spore, entering roots and establishing mycorrhizal associations.

Spores are between 30 and 500 micrometres in size - this is between 30 thousandths of a millimetre and half a millimetre. Identifying and classifying the spores is therefore something that must be done with the aid of a microscope.

Benefits of Mycorrhizal Associations

Mycorrhizal fungi are characterised by very thin hyphae, which are between 1 and 10 thousandths of a millimetre in width. These hyphae explore the soil for nutrients, transport them back to the host-plant, and help bind soil particles into aggregates. The hyphae form networks between neighbouring soil particles, between roots and soil particles, between roots on the same plant, and between roots of different plants (even different types of plants). They also form networks inside the roots they colonize. These networks of hyphae are also referred to as mycelium.

Mycorrhizas extend the volume of soil explored by the plant, a characteristic that is especially important for phosphorous which does not move in the soil solution as nitrogen does. There is some evidence that the fungi may help the plants tolerate drought. When phosphorus is scarce in soil, plants that have developed mycorrhizas on their root systems have greater access to and take up more phosphorus others. Trace elements, copper and zinc behave in a similar way to phosphorus in soil and plant roots must explore the soil to intercept them.

Although arbuscular mycorrhizas can enhance plant growth in phosphorous deficient soils, the extent to which this occurs in the field in agricultural and natural environments is difficult to measure.

Plant Pathogenic Fungi

Disease-causing microorganisms have always been inherent members of any living community. In natural ecosystems, characterised by uncontrolled and changeable conditions, their population growth is impeded by the scattered distribution of host plants and, in the case of fungal pathogens, by their dependence on rainfall at the time of spore germination. In managed systems, however, such as agriculture and horticulture, monocultures of crop plant species provide an unconstrained food supply for a pathogen. Irrigated systems also provide a constant supply of water which can enable spores to germinate and cause disease in accessible host plants.

Fungal Soil Pathogens

Climatic patterns can affect the types of fungal pathogens that are dominant in a region. For example, low fertility soils favour necrotrophic pathogens over biotrophic pathogens. Necrotrophic pathogens are distinguished from biotrophs because they kill host tissue prior to colonisation. Biotrophic pathogens include powdery mildew, downy mildew, rust, nematodes and viruses. Biotrophs live on living tissue and die when the host plant dies.

Soil Conditions Favour the Growth of Fungal Pathogens

The soil conditions that exist at the opening of the cropping season (warm-moist soils and low microbial activity) can favour the growth of a pathogen.

Reduced tillage practices help maintain infested residues at the surface of soil, increasing the damage to young seedlings. Conventional cultivations bury this inoculum source which gets broken-down more rapidly by soil microorganisms than when on the soil surface. Rotations with susceptible hosts can increase the inoculum potential of the pathogen in soil. Certain herbicides also increase the disease severity (e.g: the disease caused by the Take-all fungus and root rot caused by Rhizoctonia).

SOIL BACTERIA

Most bacteria in soil are about one micron in length or diameter (there are a thousand microns in a millimetre). Some are slightly larger than this, up to several microns, and in rare cases even larger. Their size varies with their environment. Bacteria in environments that have high levels of nutrients may be larger than those in nutrient poor conditions. The majority of bacteria in soil usually occur as single cells. Bacteria sometimes join together in chains or clusters. They mainly have one of two shapes - spheres (called cocci) and rods (called bacilli). Other bacteria have more varied shapes including spirals and long thin hyphae (although these are less common).

Importance of Bacteria in the Soil

Bacteria are able to perform an extremely wide range of chemical transformations, including degradation of organic matter, disease suppression, disease, and nutrient transformations inside roots (*e.g.,,* reducing bacteria in roots, bacteria cause nitrogen fixation).

Azobacter, for example, is a genus of free living bacteria that converts atmospheric nitrogen into ammonium, making it available for plant use. This process may only take place, however, if the following conditions are met:

- An easily degradable carbon source is available
- Any nitrogen compounds such as ammonium or nitrate, are not already in present in substantial concentrations
- Soil pH levels are between 6 and 9
- High levels of phosphorus are present
- Very low levels of oxygen are present

Azobacter is inhibited by a large range of toxic mineral and organic compounds, but may tolerate relatively high salinity and their activities are enhanced in the presence of clays. In general, bacteria are the organisms in soil that are mainly responsible for transforming inorganic constituents from one chemical form to another. Their system of external digestion means that some of the metabolites released by the use of extracellular enzymes may be used by other organisms, such as plants. The bacteria gain nutrients and energy from these processes and provide other organisms with suitable forms of chemicals they require for their own processes. For example, in the conversions of nitrate to nitrite, sulphate to sulphide and ammonium to nitrite.

Where do Bacteria Live in Soil

Bacteria are aquatic organisms that live in the water-filled pore spaces within and between soil aggregates. As such, their activities are directly dependent on relatively high soil water contents. Bacteria are normally found on the surfaces of mineral or organic particles or congregate around particles of decaying plant and animal debris. Most are unable to move and hence, their dispersion is dependent on water movement, root growth or the activity of soil and other organisms.

RHIZOBIA

Rhizobia are one of the groups of microorganisms living in soil. They are single celled bacteria, approximately one thousandth of a millimetre in length. Rhizobia belong to a family of bacteria called Rhizobiaceae. There are a number of groups (genera and species) of bacteria in this family.

Rhizobia belong to a specific group of bacteria that form a mutually beneficial association, or symbiosis, with legume plants. These bacteria take nitrogen from the air (which plants cannot use) and convert it into a form of nitrogen called ammonium (NH_4^+), which plants can use. The nitrogenase enzyme controls the process, called nitrogen fixation, and these bacteria are often called "nitrogen fixers".

Rhizobia are found in soils of many natural ecosystems. They may also be present in agricultural areas where they are associated with both crop legumes (like soybean) and pasture legumes (like clover). Usually, the rhizobia in agricultural areas have been introduced at sowing by applying an inoculum to the exterior of the seeds as liquid formations or pellets.

Nodules Formed on the Roots of Legumes

The nodulation process is a series of events in which rhizobia interact with the roots of legume plants to form a specialised structure called a root nodule. These are visible, ball-like structures that are formed by the plant in response to the presence of the bacteria.

The process involves complicated signals between the bacteria and the host roots. In the first stages, the bacteria multiply near the root and then adhere to it. The small hairs on the root's surface curl around the bacteria and they enter the root. Alternatively, the bacteria may enter directly through points on the root surface. The method of entry of the bacteria into the root depends on the type of plant. Once inside the root, the bacteria multiply within thin threads. Signals stimulate cell multiplication of both the plant's cells and the bacteria and this repeated division results in a mass of root cells containing many bacterial cells. Some of these bacteria then change into a form that is able to convert gaseous nitrogen into ammonium nitrogen (that is, they can "fix" nitrogen). These bacteria are then called bacteroids.

The shape of the nodules is controlled by the plant and nodules can vary considerably - both in size and shape. Most plants need specific kinds of rhizobia to form nodules. The rhizobia that form nodules on peas, for example, cannot form nodules on clover. Nodulation can be impeded by low pH, Al toxicity, nutrient deficiencies, salinity, waterlogging, and the presence of root parasites such as nematodes or genetic incompatibility with the plant host.

Why are Nodules Pink Inside

The nitrogenase enzyme is extremely sensitive to oxygen and is only active at low oxygen levels or anaerobic conditions. The physical structure of the nodule acts as a barrier to oxygen and the enzyme leghaemaglobin binds oxygen and transports it away from nitrogenase to respiratory sites. Leghaemoglobin gives the inside of nodules their reddish-pink colouring.

Other Plant-bacterial Associations that Fix Atmospheric Nitrogen

Associations between bacteria and plants that fix atmospheric nitrogen include an association between species of Frankia bacteria and several tree species such as those of the genera Casuarina and Allocasuarina. Another example is between that of Azospirillum and grasses.

CONTROLLING THE SOIL MICROFLORA

PRINCIPLES AND STRATEGIES

Principles of Natural Ecosystems and the Application of Beneficial and Effective Microorganisms. The misuse and excessive use of chemical fertilizers and pesticides have often adversely affected the environment and created many a) food safety and quality and b) human and animal health problems. Consequently, there has been a growing interest in nature farming and organic agriculture by consumers and environmentalists as possible alternatives to chemical-based, conventional agriculture. Agricultural systems which conform to the principles of natural ecosystems are now receiving a great deal of attention in both developed and developing countries. A number of books and journals have recently been published which deal with many aspects of natural farming systems. New concepts such as alternative agriculture, sustainable agriculture, soil quality, integrated pest management, integrated nutrient management and even beneficial microorganisms are being explored by the agricultural research establishment. Although these concepts and associated methodologies hold considerable promise, they also have limitations. For example, the main limitation in using microbial inoculants is the problem of reproducibility and lack of consistent results.

Unfortunately certain microbial cultures have been promoted by their suppliers as being effective for controlling a wide range of soil-borne plant diseases when in fact they were effective only on specific pathogens under very specific conditions. Some suppliers have suggested that their particular microbial inoculant is akin to a pesticide that would suppress the general soil microbial population while increasing the population of a specific beneficial microorganism. Nevertheless, most of the claims for these single-culture microbial inoculants are greatly exaggerated and have not proven to be effective under field conditions. One might speculate that if all of the microbial cultures and inoculants that are available as marketed products were used some degree of success might be achieved because of the increased diversity of the soil microflora and stability that is associated with mixed cultures. While this, of course, is a hypothetical example, the fact remains that there is a greater likelihood of controlling the soil microflora by introducing mixed, compatible cultures rather than single pure cultures.

Even so, the use of mixed cultures in this approach has been criticized because it is difficult to demonstrate conclusively which microorganisms are responsible for the observed effects, how the introduced microorganisms interact with the indigenous species, and how these new associations affect the soil/plant environment. Thus, the use of mixed cultures of beneficial microorganisms as soil inoculants to enhance the growth, health, yield, and quality of crops has not gained widespread acceptance by the agricultural research establishment because conclusive scientific proof is often lacking.

The use of mixed cultures of beneficial micr oorganisms as soil inoculants is based on the principles of natural ecosystems which are sustained by their constituents; that is, by the quality and quantity of their inhabitants and specific ecological parameters, *i.e.*, the greater the diversity and number of the inhabitants, the higher the order of their interaction and the more stable the ecosystem. The mixed culture approach is simply an effort to apply these principles to natural systems such as agricultural soils, and to shift the microbiological equilibrium in favour of increased plant growth, production and protection.

It is important to recognize that soils can vary tremendously as to their types and numbers of microor-ganisms. These can be both beneficial and harmful to plants and often the predominance of either one depends on the cultural and management practices that are applied. It should also be emphasized that most fertile and productive soils have a high content of organic matter and, generally, have large, populations of highly diverse microorganisms (*i.e.*, both species and genetic diversity). Such soils will also usually have a wide ratio of beneficial to harmful microorganisms.

CONTROLLING THE SOIL MICROFLORA FOR OPTIMUM CROP PRODUCTION AND PROTECTION

The idea of controlling and manipulating the soil microflora through the use of inoculants organic amendments and cultural and management practices to create a more favourable soil microbiological environment for optimum crop production and protection is not new. For almost a century, microbiologists have known that organic wastes and residues, including animal manures, crop residues, green manures, municipal wastes (both raw and composted), contain their own indigenous populations of microorganisms often with broad physiological capabilities.

It is also known that when such organic wastes and residues are applied to soils many of these introduced microorganisms can function as biocontrol agents by controlling or suppressing soil-borne plant pathogens through their competitive and antagonistic activities. While this has been the theoretical basis for controlling the soil microflora, in actual practice the results have been unpredictable and inconsistent, and the role of specific microorganisms has not been well-defined.

For, many years microbiologists have tried to culture beneficial microorganisms for use as soil inoculants to overcome the harmful effects of phytopathogenic organisms, including bacteria, fungi and nematodes. Such attempts have usually involved single applications of pure cultures of microorganisms which have been largely unsuccessful for several reasons. First, it is necessary to thoroughly understand the individual growth and survival characteristics of each particular beneficial micro organism, including their nutritional and environmental requirements. Second, we must understand their ecological relationships and interactions with other microorganisms, including their ability to coexist in mixed cultures and after application to soils.

There are other problems and constraints that have been major obstacles to controlling the microflora of agricultural soils. First and foremost is the large number of types of microorganisms that are present at any one time, their wide range of physiological capabilities, and the dramatic fluctuations in their populations that can result from man's cultural and management practices applied to a particular farming system. The diversity of the total soil microflora depends on the nature of the soil environment and those factors which affect the growth and activity of each individual organism including temperature, light, aeration, nutrients, organic matter, pH and water. While there are many microorganisms that respond positively to these factors, or a combination thereof, there are many that do not. Microbiologists have actually studied relatively few of the microorganisms that exist in most agricultural soil, mainly because we don't know how to culture them; *i.e.*, we know very little about their growth, nutritional, and ecological requirements.

The "diversity" and "population" factors associated with the soil microflora have discouraged scientists from conducting research to develop control strategies. Many believe that, even when beneficial microorganisms are cultured and inoculated into soils, their number is relatively small compared with the indigenous soil inhabitants, and they would likely be rapidly overwhelmed by the established soil microflora. Consequently, many would argue that even if the application of beneficial microorganisms is successful under limited conditions (*e.g.*, in the laboratory) it would be virtually impossible to achieve the same success under actual field conditions. Such thinking still exists today, and serves as a principle constraint to the concept of controlling the soil microflora.

It is noteworthy that most of the microorganisms encountered in any particular soil are harmless to plants with only a relatively few that function as plant pathogens or potential pathogens. Harmful microorganisms become dominant if conditions develop that are favourable to their growth, activity and reproduction. Under such conditions, soil-borne pathogens (*e.g.*, fungal pathogens) can rapidly increase their populations with devastating effects on the crop. If these conditions change, the pathogen population declines just as rapidly to its original state. Conventional farming systems that tend towards the consecutive planting of the same crop (*i.e.*, monoculture) necessitate the heavy use of chemical fertilizers and pesticides. This, in turn, generally increases the probability that harmful, disease-producing, plant pathogenic microorganisms will become more dominant in agricultural soils.

Chemical-based conventional farming methods are not unlike symptomatic therapy. Examples of this are applying fertilizers when crops show symptoms of nutrient-deficiencies, and applying pesticides whenever crops are attacked by insects and diseases. In efforts to control the soil microflora some scientists feel that the introduction of beneficial microorganisms should follow a symptomatic approach. However, we do not agree. The actual soil conditions that prevail at any point in time may be most unfavourable to the growth and establishment of laboratory-cultured, beneficial microorganisms.

To facilitate their establishment, it may require that the farmer make certain changes in his cultural and management practices to induce conditions that will:

- Allow the growth and survival of the inoculated microorganisms and
- Suppress the growth and activity of the indigenous plant pathogenic microorganisms.

An example of the importance of controlling the soil microflora and how certain cultural and management practices can facilitate such control is useful here. Vegetable cultivars are often selected on their ability to grow and produce over a wide range of temperatures. Under cool, temperate conditions there are generally few pest and disease problems. However, with the onset of hot weather, there is a concomitant increase in the incidence of diseases and insects making it rather difficult to obtain acceptable yields without applying pesticides.

With higher temperatures, the total soil microbial population increases as does certain plant pathogens such as Fusarium, which is one of the main putrefactive, fungal pathogens in soil. The incidence and destructive activity of this pathogen can be greatly minimized by adopting reduced tillage methods and by shading techniques to keep the soil cool during hot weather.

Another approach is to inoculate the soil with beneficial, antagonistic, antibiotic-producing microorganisms such as actinomycetes and certain fungi.

APPLICATION OF BENEFICIAL AND EFFECTIVE MICROORGANISMS:A NEW DIMENSION

Many microbiologists believe that the total number of soil microorganisms can be increased by applying organic amendments to the soil. This is generally true because most soil microorganisms are heterotrophic, *i.e.*, they require complex organic molecules of carbon and nitrogen for metabolism and biosynthesis. Whether the regular addition of organic wastes and residues will greatly increase the number of beneficial soil microorganisms in a short period of time is questionable. However, we do know that heavy applications of organic materials, such as seaweed, fish meal, and chitin from crushed crab shells, not only helps to balance the micronutrient content of a soil but also increases the population of beneficial antibiotic-producing actinomycetes. This changes the soil to a disease-suppressive condition within a relatively short period.

The probability that a particular beneficial microorganism will become predominant, even with organic farming or nature farming methods, will depend on the ecosystem and environmental conditions. It can take several hundred years for various species of higher and lower plants to interact and develop into a definable and stable ecosystem. Even if the population of a specific microorganism is increased through cultural and management practices, whether it will be beneficial to plants is another question. Thus, the likelihood of a beneficial, plant-associated microorganism becoming predominant under conservation-based farming systems is virtually impossible to predict. Moreover, it is very unlikely that the population of useful anaerobic microorganisms, which usually comprise only a small part of the soil microflora, would increase significantly even under natural farming conditions.

This information then emphasizes the need to develop methods for isolating and selecting different microor ganisms for their beneficial effects on soils and plants. The ultimate goal is to select microorganisms that are physiologically and ecologically compatible with each other and that can be introduced as mixed cultures into soil where their beneficial effects can be realised.

APPLICATION OF BENEFICIAL AND EFFECTIVE MICROORGANISMS: FUNDAMENTAL CONSIDERATIONS

Microorganisms are utilized in agriculture for various purposes; as important components of organic amen dments and composts, as legume inoculants for biological nitrogen fixation as a means of suppressing insects and plant diseases to Improve crop quality and yields, and for reduction of labour. All of these are closely related to each other. An important consideration in the application of beneficial microorganisms to soils is the enhancement of their synergistic effects. This is difficult to accomplish if these microorganisms are applied to achieve symptomatic therapy, as in the case of chemical fertilizers and pesticides.

If cultures of beneficial microorganisms are to be effective after inoculation into soil, it is important that their initial populations be at a certain critical threshold level. This helps to ensure that the amount of bioactive substances produced by them will be sufficient to achieve the desired positive effects on crop production and/or crop protection. If these conditions are not met, the introduced microorganisms, no matter how useful they are, will have little if any effect. At present, there are no chemical tests that can predict the probability of a particular soil-inoculated microorganism to achieve a desired result. The most reliable approach is to inoculate the beneficial microorganism into soil as part of a mixed culture, and at a sufficiently high inoculum density to maximize the probability of its adaptation to enviro nmental and ecological conditions.

The application of beneficial microorganisms to soil can help to define the structure and establishment of natural ecosystems. The greater the diversity of the cultivated plants that are grown and the more chemically complex the biomass, the greater the diversity of the soil microflora as to their types, numbers and activities.

The application of a wide range of different organic amen dments to soils can also help to ensure a greater microbial diversity. For example, combinations of various crop residues, animal manures, green manures, and municipal wastes applied periodically to soil will provide a higher level of microbial diversity than when only one of these materials is applied. The reason for this is that each of these organic materials has its own unique indigenous microflora which can greatly affect the resident soil microflora after they are applied, at least for a limited period.

CLASSIFICATION OF SOILS BASED ON THEIR MICROBIOLOGICAL PROPERTIES

Most soils are classified on the basis of their chemical and physical properties; little has been done to classify soils according to their physicochemical and microb iological properties. The reason for this is that a soil's chemical and physical properties are more readily defined and measured than their microbiological properties.

Improved soil quality is usually characterized by increased infiltration; aeration, aggregation and organic matter content and by decreased bulk density, compaction, erosion and crusting. While these are important indicators of potential soil productivity, we must give more attention to soil biological properties because of their important relationship (though poorly understood) to crop production, plant and animal health, environmental quality, and food safety and quality. Research is needed to identify and quantify reliable and predictable biological/ecological indicators of soil quality. Possible indicators might include total species diversity or genetic diversity of beneficial soil microorganisms as well as insects and animals.

The basic concept here is not to classify soils for the study of microorganisms but for farmers to be able to control the soil microflora so that biologically-mediated processes can improve the growth, yield, and quality of crops as well as the tilth, fertility, and productivity of soils. The ultimate objective is to reduce the need for chemical fertilizers and pesticides.

FUNCTIONS OF MICROORGANISMS. PUTREFACTION, FERMENTATION, AND SYNTHESIS

Soil microorganisms can be classified into decomposer and synthetic microorganisms. The decomposer microorganisms are subdivided into groups that perform oxidative and fermentative decomposition. The fermentative group is further divided into useful fermentation (simply called fermentation) and harmful fermentation (called putrefaction). The synthetic microorganisms can be sub-divided into groups having the physiological abilities to fix atmospheric nitrogen into amino acids and/or carbon dioxide into simple organic molecules through photosynthesis.

Fermentation is an anaerobic process by which facultative microorganisms (*e.g.*, yeasts) transform complex organic molecules (*e.g.*, carbohydrates) into simple organic compounds that often can be absorbed directly by plants. Fermentation yields a relatively small amount of energy compared with aerobic decomposition of the same substrate by the same group of microorganisms.

Aerobic decomposition results in complete oxidation of a substrate and the release of large amounts of energy, gas, and heat with carbon dioxide and water as the end products. Putrefaction is the process by which facultative heterotrophic microorganisms decompose proteins anaerobically, yielding malodorous incompletely oxidized, metabolites (*e.g.*, ammonia, mercaptans and indole) that are often toxic to plants and animals.

The term "synthesis" as used here refers to the biosynthetic capacity of certain microorganisms to derive metabolic energy by "fixing" atmospheric nitrogen and/or carbon dioxide. In this context we refer to these as "synthetic" microorganisms, and if they should become a predominant part of the soil microflora, then the soil would be termed a "synthetic" soil.

Nitrogen-fixing microorganisms are highly diverse, ranging from "free-living" autotrophic bacteria of the genus Azotobacter to symbiotic, heterotrophic bacteria of the genus Rhizobium, and blue-green algae (now mainly classified as blue-green bacteria), all of which function aerobically. Photosynthetic microorganisms fix atmospheric carbon dioxide in a manner similar to that of green plants. They are also highly diverse, ranging from blue-green algae to green algae that perform complete photosynthesis aerobically to photosynthetic bacteria which perform incomplete photosynthesis anaerobically.

RELATIONSHIPS BETWEEN PUTREFACTION, FERMENTATION, AND SYNTHESIS

The processes of putrefaction, fermentation, and synthesis proceed simultaneously according to the appropriate types and numbers of microorganisms that are present in the soil. The impact on soil quality attributes and related soil properties is determined by the dominant process. The production of organic substances by microorganisms results from the intake of positive ions, while decomposition serves to release these positive ions. Hydrogen ions play a pivotal role in these processes.

A problem occurs when hydrogen ions do not recombine with oxygen to form water but are utilized to produce methane, hydrogen sulfide, ammonia, mercaptans and other highly reduced putrefactive substances most of which are toxic to plants and produce malodors. If a soil is able to absorb the excess hydrogen ions during periods of soil anaerobiosis and if synthetic microorganisms such as photosynthetic bacteria are present, they will utilize these putrefactive substances and produce useful substrates from them which helps to maintain a healthy and productive soil.

The photosynthetic bacteria, which perform incomplete photosynthesis anaerobically, are highly desirable, beneficial soil microorganisms because they are able to detoxify soils by transforming reduced, putrefactive substances such as hydrogen sulfide into useful substrates. This helps to ensure efficient utilization of organic matter and to improve soil fertility. Photosynthesis involves the photo-catalyzed splitting of water which yields molecular oxygen as a by-product. Thus, these microorganisms help to provide a vital source of oxygen to plant roots.

Reduced compounds such as methane and hydrogen sulfide are often produced when organic materials are decomposed under anaerobic conditions. These compo unds are toxic and can greatly suppress the activities of nitrogen-fixing microorganisms. However, if synthetic microorganisms, such as photosynthetic bacteria that utilize reduced substances, are present in the soil, oxygen deficiencies are not likely to occur. Thus, nitrogen-fixing microorganisms, coexisting in the soil with photosy nthetic bacteria, can function effectively in fixing atmospheric nitrogen even under anaerobic conditions.

Photosynthetic bacteria not only perform photos ynthesis but can also fix-nitrogen. Moreover, it has been shown that, when they coexist, in soil with species of Azotobacter, their ability to fix nitrogen is enhanced. This then is an example of a synthetic soil. It also suggests that by recognizing the role, function, and mutual compatibility of these two bacteria and utilizing them effectively to their full potential, soils can be induced to a greater synthetic capacity. Perhaps the most effective synthetic soil system results from the enhancement of zymogenic and synthetic microorganisms; this allows fermentation to become dominant over putrefaction and useful synthetic processes to proceed.

CLASSIFICATION OF SOILS BASED ON THE FUNCTIONS OF MICROORGANISMS

As discussed earlier, soils can be characterized according to their indigenous microflora which perform putrefactive, fermentative, synthetic and zymogenic reactions and processes. In most soils, these three functions are going on simultaneously with the rate and extent of each determined by the types and numbers of associated microorganisms that are actively involved at any one time.

Disease-Inducing Soils. In this type of soil, plant pathogenic microorganisms such as Fusarium fungi can comprise 5 to 20 per cent of the total microflora if fresh organic matter with a high nitrogen content is applied to such a soil, incompletely oxidized products can arise that are malodorous and toxic to growing plants. Such soils tend to cause frequent infestations of disease organisms, and harmful insects. Thus, the application of fresh organic matter to these soils is often harmful to crops. Probably more than 90 per cent of the agricultural land devoted to crop production worldwide can be classified as having disease-inducing soil. Such soils generally have poor physical properties, and large amounts of energy are lost as "greenhouse" gases, particularly in the case of rice fields. Plant nutrients are also subject to immobilization into unavailable forms.

Disease-Suppressive Soils

The microflora of disease-suppressive soils is usually dominated by antagonistic microorganisms that produce copious amounts of antibiotics. These include fungi of the genera Penicillium, Trichoderma, and Aspergillus, and actinomycetes of the genus Streptomyces. The antibiotics they produce can have biostatic and biocidal effects on soil-borne plant pathogens, including Fusarium which would have an incidence in these soils of less than 5 per cent. Crops planted in these soils are rarely affected by diseases or insect pests. Even if fresh organic matter with a high nitrogen content is applied, the production of putrescent substances is very low and the soil has a pleasant earthy odor after the organic matter is decomposed.

These soils generally have excellent physical properties; for example, they readily, form water-stable aggregates and they are well-aerated, and have a high permeability to both air and water. Crop yields in the disease-suppressive soils are often slightly lower than those in synthetic soils. Highly acceptable crop yields are obtained whenever a soil has a predominance of both disease-suppressive and synthetic microorganisms.

Zymogenic Soils

These soils are dominated by a microflora that can perform useful kinds of fermentations, *i.e.*, the breakdown of complex organic molecules into simple organic substances and inorganic materials. The organisms can be either obligate or facultative anaerobes. Such fermentation-producing microorganisms often comprise the microflora of various organic materials, *i.e.*, crop residues, animal manures, green manures and municipal wastes including composts. After these amendments are applied to the soil, their number: and fermentative activities can increase dramatically and overwhelm the indigenous soil microflora for an indefinite period.

While these microorganisms remain predominant, the soil can be classified as a zymogenic soil which is generally characterized by,

- Pleasant, fermentative odors especially after tillage,
- Favourable soil physical properties (*e.g.*, Increased aggregate stability, permeability, aeration and decreased resistance to tillage
- Large amounts of inorganic nutrients, amino acids, carbohydrates, vitamins and other bioactive substances which can directly or indirectly enhance the growth, yield and quality of crops,
- Low occupancy of Fusarium fungi which is usually less than 5 per cent, and
- Low production of greenhouse gases (*e.g.*, methane, ammonia, and carbon dioxide) from croplands, even where flooded rice is grown.

Synthetic Soils

These soils contain significant populations of microorganisms which are able to fix atmospheric nitrogen and carbon dioxide into complex molecules such as amino acids, proteins and carbohydrates. Such microorganisms include photosynthetic bacteria which perform incomplete photosynthesis anaerobically, certain Phycomycetes (fungi that resemble algae), and both green algae and blue—green algae which function aerobically. All of these are photosynthetic organisms that fix atmospheric nitrogen. If the water content of these soils is stable, their fertility can be largely maintained by regular additions of only small amounts of organic materials. These soils have a low Fusarium occupancy and they are often of the disease-suppressive type. The production of gases from fields where synthetic soils are present is minimal, even for flooded rice.

This is a somewhat simplistic classification of soils based on the functions of their predominant types of microorganisms, and whether they are potentially beneficial or harmful to the growth and yield of crops. While these different types of soils are described here in a rather idealized manner, the fact is that in nature they are not always clearly defined because they often tend to have some of the same characteristics. Nevertheless, research has shown that a disease-inducing soil can be transformed into disease-suppressing, zymogenic and synthetic soils by inoculating the problem soil with mixed cultures of effective microorganisms. Thus it is somewhat obvious that the most desirable agricultural soil for optimum growth, production, protection, and quality of crops would be the composite soil indicated in Fig., *i.e.*, a soil that is highly zymogenic and synthetic, and has an established disease-suppressive capacity. This then is the principle reason for seeking ways and means of controlling the microflora of agricultural soils.

PESTICIDE TREATMENT ON BACTERIAL SOIL

At least once a year about 85 per cent of agriculturally used soils in western Europe are treated with herbicides in order to ensure that high-quality agricultural goods will be produced at levels that cover costs. Actual registration procedures include evaluation of the impact of herbicides on the environment by testing for non-target effects on single species or microbial communities.

The impact on soil microbial communities is evaluated in view of their role in sustaining the global cycling of matter and their varied functions in supporting plant growth. In order to observe effects by quantifying changes in microbial biomass or in the general metabolic activities of microbial communities, a variety of laboratory tests have been recommended. Internationally, there is wide divergence in the extent to which the performance of such tests is obligatory for registrations. Substrate-induced-respiration (SIR), dehydrogenase activity, and nitrogen turnover are investigated in German registration procedures.

The traditional techniques used in our study can be divided into two groups. SIR and dehydrogenase activity are both strictly correlated to biomass, whereas long-term respiration and nitrogen mineralization are not. The immediate respiration of a microbial community following a glucose addition is quantified in a manner avoiding a significant contribution of cell multiplication. Testing for the activity of dehydrogenases, as determined by the accumulation of reduced triphenyl formazane, conveniently monitors the metabolic activities of microbial cells in oxidizing organic substrates. This parameter seems to indicate an impact on microbial activity most significantly. The efficiency of microbial carbon mineralization over an extended period (8 weeks in this report) can be stimulated by the mineralization of dead cells or the degradation of the pesticide itself. Carbon mineralization and nitrogen mineralization serve to complement the interpretation of the results for the biomass-related parameters.

Another approach to differentiating bacterial communities by their catabolic potentials is offered by the BIOLOG system (BIOLOG Inc., Hayward, Calif.), originally commercialized for the identification of bacterial strains. The metabolism of 95 carbon sources in microtiter plates, indicated by the reduction of a tetrazolium dye, affords a detailed view of the catabolic activities of complex bacterial communities. These features seem to be suited to monitoring of the impact of stress factors on microbial communities. In contrast to such approaches, testing for effects on a special activity of a single microbial species can be used in some cases of soil contamination. Such testing may indicate how the whole community responds to toxic interactions but might be considered to be restricted in its ecological significance.

Analysis of changes in the composition of a community can also be used to characterize its response to stress. Species abundance and diversity measurements are sensitive indicators of environmental conditions. However, the conventional microbiological investigation of community diversity has significant pitfalls. First, the time and resources needed for quantification and identification of morphologically similar microbial species preclude an extensive analysis of many samples from different treatments. The counting of colony morphotypes has not been tested as an alternative on a broader scale. Second, a significant bias invariably is introduced by the selective process of cultivation. Microbiologists are becoming increasingly aware of the fact that only a small fraction of microorganisms is analyzed by conventional cultivation methods, and frequently these species do not seem to represent the spectrum of organisms active in the habitat of choice.

A measurement more representative of the structure of communities in situ may be obtained by isolating and analyzing macromolecules of microbial origin from a habitat, as in other methods of microbial-community characterization. A community's composition and diversity are most significantly represented by the complexity of information in nucleic acid molecules. A great variety of methods of isolating DNA and RNA from different habitats, either by direct extraction or following a prior separation of microbial cells, have been published. The information in about 4,000 different bacterial chromosomes, equivalent to more than 10,000 species, seems to represent the complexity of microbial communities in healthy agricultural soils, as estimated by DNA reassociation kinetics. Wide divergence in levels of diversity among different soils can be demonstrated, but the technique does not seem to be adaptable to a measurement of induced changes, particularly with respect to its sensitivity.

A more detailed view of the composition of a community can be obtained by sampling the nucleic acid population from a habitat in order to determine the abundance of phylogenetically meaningful sequences, particularly the 16S rRNA and 16S rDNA. The origins of individual sequences are then classified by comparison to large databases and their present interpretation. The frequency of sequences within the sample represents the community composition in some detail, depending on the number of sequences analyzed. Such approaches have become an established method of molecular ecology. Unfortunately, these techniques are labour intensive and time-consuming, a major drawback for any investigation involving a large number of differently treated samples.

The sequence-dependent separation of 16S rDNA fragments in a denaturing gradient gel (denaturing gradient gel electrophoresis [DGGE]) in order to analyze the abundance of divergent sequences from particular communities was introduced by Muyzer *et al.* By superimposing a temperature gradient instead of a chemical gradient on the electric field, temperature gradient gel electrophoresis (TGGE) applies the identical principle of separating double-stranded DNA fragments.

Minor sequence variations, including point mutations, result in modified migration and distinct bands.

The technique of generating a pattern from sequence variants with well-defined phylogenetic meaning is attractive from different points of views. The pattern yields a representation of the community as a whole or of a section of it, defined by the selected primers. The bands of the pattern are not directly interpretable in phylogenetic terms. Their "fingerprint," however, represents the frequency distribution of different sequences in a sample. Variations in such fingerprints due to induced variations in community composition may be easily observed. The degrees of pattern change may be quantified to some extent to allow for ranking. Finally, band isolation and sequencing is one straightforward possibility to investigate induced changes in detail with respect to responsive community members.

We have used both the BIOLOG system and pattern generation by amplified ribosomal sequences to investigate the suitability of these approaches for the analysis of the impacts of herbicide applications on the function and structure of soil bacterial communities. The more traditional methods for analysis of such impacts serve as a well-defined background of ecotoxicological testing.

TYPES OF SIDE-EFFECTS ON BENEFICIAL ORGANISMS

Pesticides can exhibit primary or secondary effects on predators, parasitoids and pathogens of target pests. Primary effects are direct or indirect, depending on their exposure and on the biological parameter influenced. Direct mortality of beneficial organisms may be caused by direct contact during application, pesticide residues, taking up contaminated prey, intoxication by fumigants, and contact or contamination with soil disinfectants. Indirect or sublethal effects on beneficial arthropods include decreases in reproduction, oviposition, parasitization, predation, longevity and egg viability, and a delay in development and shifting of the sex-ratio. Morphological and behavioural changes may also occur. Secondary effects due to pesticides include killing the prey/host of a beneficial organism or of species which produce alternative food like honeydew, taking up contaminated food, and directly stimulating the pest; for example, some pyrethroids enhance reproduction in *Tetranychus urticae* Koch. Pesticides directly affect entomopathogenic fungal biocontrol agents by inhibition of spore germination and vegetative development (mycelial growth), and they also reduce the viability of conidia and their survival and activity on plant surfaces. Viability and infectivity of the infective juveniles (J3) of entomopathogenic nematodes are also adversely affected. Side-effects of pesticides on natural enemies may vary between and within taxonomic groups. From their comprehensive data on the side-effects of pesticides, Theiling and Croft concluded that predators were more tolerant to pesticide treatment than parasitoids, except for fungicides, towards which susceptibility was not greatly affected.

The tolerance of aphid natural enemies decreases from Coccinellids > Chrysopids > Syrphids > Hemiptera > Hymenoptera. Evaluation of effects within taxonomic groups revealed that the classification of the effects of 74 compounds tested against the parasitoids *Encarsia formosa* Gahan, *Aphidius matricariae* Haliday and *Leptomastix dactylopii* Howard corresponded by more than 78 per cent. In a comparison of trial results with 81 test compounds for predatory mite species occurring in orchards and vineyards with *Phyloseiulus persimilis* Athias-Henriot, the same level was reached in 64 per cent of the test compounds.

Differences in susceptibility have been recorded between taxonomically close species, and even between strains within the same species. *Eretmocerus mundus* Mercet adults were less susceptible to residues of amitraz, thiodicarb and cypermethrin than *E. formosa* or *Encarsia pergandiella* Howard. Among *Aphidius* species, *A. matricariae* was more tolerant to dimethoate than *Aphidius rhopalosiphi* de Stefani Perez or *Aphidius colemani* Viereck. The response of several species of entomopathogenic fungi to copper incorporated in agar differed. *Paecilomyces farinosus* (Holmsk.) A.H.S. Brown and G. Sm. was more tolerant than *Verticilium lecanii* (A. Zimmerm.) Viégas, *Beauveria bassiana* (Balsamo) Vuillemin and *Metarhizium anisopliae* (Metschnikoff) Sorokin. The entomopathogenic nematodes *Steinemema carpocapsae* (Weiser), *Steinemema feltiae* (Filipjev) and *Heterorhabditis* HP88 exhibited different tolerance levels to 9 tested pesticides. Repeated exposure of local strains to chemicals may cause natural enemies to develop tolerance to pesticides.

This is the case of *P. persimilis* and organophosphorous compounds (OPs) and of *Aphidoletes aphidimyza* (Rondani) and azinphos-methyl. Developmental stage may greatly influence the response of natural enemies to pesticides. The susceptibility of *A. aphidimyza* and *Chrysoperla cornea* (Stephens) to pesticides with contact mode of action increased from the egg stage to the adults. In contrast, pesticide susceptibility was lowest in treated adults of *Coccinella septempunctata* L. and in eggs of *P. persimilis,* while in the coccinellid the egg stage and in the predatory mite the larvae or protonymph stage were the least tolerant. However, compounds with modes of action that regulate or inhibit insect growth resulted in high mortality of *C. carnca* larvae, but not of the adults, whose fertility was only slightly affected.

The host may offer parasitoids different degrees of protection against pesticides; unprotected stages of parasitoids (*e.g.* adult hymenoptera) and protected stages (*e.g.* different developmental stages in aphid mummies) show different levels of mortality after the same pesticide treatment. Avermectin B killed 50 per cent of *E. formosa* protected in the whitefly scales in a direct contact test, but 79 per cent of the adult wasps after contact with the dried residue. *Leptomastix dactylopii* protected in *Planococcus citri* (Risso) were barely affected by topical treatment of endosulfan, while the adults were severely damaged in residual tests. Even sexes of the same species may present different susceptibility against pesticides. In 5 different populations of *Diglyphus begini* (Ashmead) and in predatory mites, males are less tolerant than females.

Tests and Approaches to Detect Side-Effects of Pesticides

One of the most comprehensive programmes to test side-effects of pesticides on beneficial organisms was set up by the IOBC/WPRS working group "Pesticides and Beneficial Organisms". In the first step, arthropod species and microorganisms that were regarded as the most important natural enemies in the different crops were identified. For these species test methods at different levels were developed. Pesticide screening is based on a sequence of three steps in laboratory, semi-field and field conditions, as shown in figure. The sequential programme assumes that pesticides that are harmless in the laboratory will also be safe in semi-field and field conditions, and do not need to be evaluated in further steps.

When a chemical, however, is categorized as harmful in one step, its effect at the next step cannot be inferred, and the sequence must be continued until it finishes at field conditions or displays no negative effects. The pesticides are usually tested at the highest recommended field rate as commercial formulations. The laboratory methods aim to evaluate the direct, initial toxicity of pesticide residues on susceptible and protected developmental stages of the test arthropods and are thus classified as lab-a- and lab-b-tests. The aim of the first test is the detection of pesticides which are harmless to the test organism after worst case exposure to dried pesticide residue on a defined test surface (glass or sand) after a single application of the test compound. The results of the tests should include the mortality (direct effect) and the reproduction (sublethal effect) of the test organism.

Information about the duration of the effect of a pesticide is provided by the persistence test. Plant material (*e.g.* leaves) is sprayed with the test pesticide and left on the plant under greenhouse conditions for residue aging. Leaf samples undergo a further test, similar to the lab-a-test. The next test is the semi-field test which is carried out on pesticide residues or as a direct application on the plants with the test arthropods, and is kept under more natural conditions. Sublethal effects, behavioural changes, and the effect of more than one application of the test product are thus evaluated. The range of tests developed for a selection of organisms important in greenhouse crops is presented in Table.

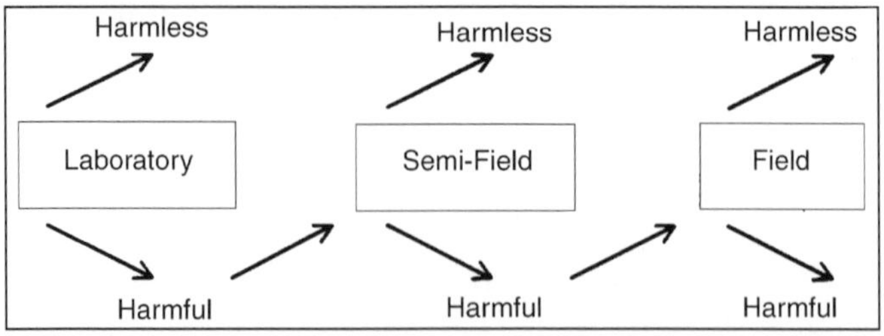

The lab-a-test for parasitoids (*E. formosa, A. matricariae*) and for A. *aphidimyza* adults (on leaf material) is carried out as a residual contact test with adult wasps or gall midges. Mortality and reproduction (parasitization of the host or number of eggs deposited) are evaluated. In the lab-b-test, the protected stages of the parasitoids in their hosts (aphid mummies; whitefly scales) are directly sprayed with the pesticide solution and the emergence rate from the hosts is assessed. The lab-a-test for predatory mites is a residual contact test starting with predatory larvae or protonymphs. During the test the mortality rate, escaping rate and reproduction per female are evaluated.

The same testing procedure is used as in the lab-a-test for *Orius niger* (Wolff), and the emergence from the deposited eggs is also assessed. The lab-b-test for *O. niger* is the same as the lab-a-test, but uses predatory bug adults. The lab-b-test for *A. aphidimyza* is carried out with larvae as a residual contact test on leaves and is also appropriate for a persistence test. Laboratory tests for *C. carnea* and *Syrphus corollae* Fabricius follow the same principles.

Larvae are tested in a residual contact test to assess the mortality and reproduction of the test organisms. A laboratory test for Coccinellids has also been described in detail for *Hippodamia oculata* (Thunberg). A residual contact test with larval stages is carried out to evaluate mortality and duration of development. The adults deriving from this first testing phase are used to check reproduction, duration of sexual maturation of females, and emergence from the deposited eggs.

Table. Test Methods on Different Test Levels of the Sequential Testing Scheme Developed within the IOBC/WPRS Working Group "Pesticides and Beneficial Organisms"

Test Organisms	Lab-a-	Lab-b-	Extended	Semi-Field	Semi-Field	Test	Test
Laboratory	Test Initial		Test				
			Test/	Tox.	Persistence/		
	Persistence			Greenhouse			
Parasitoids							
Encarsia	X	X		X	X		
Formosa							

Aphidius Matricarioe	X	X	X		
Leptomastix Dactylopii	X				
Aphytis Melinus	X				
Predatory Mites					
Phytoseiulus Persimilis	X		X	X	
Others	X		X		
Predators					
Chrysoperla Carnea	X		X	X	X
Syrphus Corollae	X				
Episyrphus Balteatus				X	X
Coccinellids (Semiadalia, Harmonia, Coccinella)	X				
Contd...					
Orius Niger	X	X	X	X	
Aphidoletes Aphidimyza	X	X	X		
Entomopathogenic Fungi					
Beauveria Bassiana, Beauveria Brongiartii	X	X	X	X	
Metarhizium Anisopliae	X	X	X	X	
Verticillium Lecanii	X				
Entomopathogenic Nematodes					
Steinernema Sp.	X				

Persistence tests or tests to detect the duration of harmful effects of the pesticide residue are very similar for nearly all test organisms. Suitable plants are sprayed and kept under greenhouse conditions for different periods. Leaf samples are collected at regular intervals and are used as test surfaces, as in the lab-a-test or the lab-b-test. Mortality and reproduction are again assessed. In this case persistence must be considered as an extended laboratory test. For *C. carnea* and *Episyrphus balteatus* (DeGeer) the test is carried out on treated plants and, in addition to the above mentioned parameters, changes in the behaviour of *E. balteatus* can be examined.

The sequential IOBC testing scheme for *B. bassiana* and *M. anisopliae* comprises all three test levels. In the lab-tests the mycelial growth on agar containing pesticides is measured. The production and viability of conidia is assessed with a bioassay to check virulence. It has been proposed to switch from tests on solid medium to a worst case test for growth inhibition in liquid medium, where the mycelium, as the most sensitive stage of the fungus, is immersed into the pesticide solution. In the semi-field test conidia are mixed with standard soil and treated with the test pesticide. The soil is then incubated and the number of spores per unit of soil is determined. To check the virulence of the tested fungus at each step of the sequential scheme the *Galleria*-bait-method may be used. The results of an *in vivo* assay, in which leaf discs are sprayed with the conidial suspension of the beneficial fungus on a dried residue of the test pesticide have been described. Side effect testing at the infective juvenile J3 stage of entomopathogenic nematodes is carried out in a 2-step scheme. First, the viability and the behaviour *in vitro* in pesticide solutions is checked. In the next step, mobility and infectivity are examined in a bioassay in soil. Compatibility of pesticides with bumble-bees used as natural pollinators in greenhouses is classified in four categories, which allow or exclude the use of bumblebees or recommend a certain period after pesticide application during which the hives should be removed from the greenhouses. Comprehensive data collections about side-effects of pesticides on natural enemies are available from commercial suppliers of beneficial organisms and also in the tables published by the IOBC/WPRS working group.

Effects of Chemical Pesticides on Beneficial Organisms Used in Greenhouses

Information on specific pesticide effects on natural enemies and pathogens may be found in the published results of the Joint Testing Programmes by the IOBC/WPRS Working Group "Pesticides and Beneficial Organisms" and in many other references.

Generally herbicides, acaricides and fungicides have less effect than insecticides, although mycopesticides are highly susceptible to fungicides:

- *Effect on beneficial predators*: For predatory mites most pyrethroids and carbamates were harmful, both in initial toxicity and in reproduction and persistence trials with the susceptible juvenile predators. *Aphidoletes aphidimyza* showed a similar susceptibility to insecticide/acaricide treatments, and was also affected by OPs. OPs caused varying levels of mortality in predatory mites. In coccinellids, high mortality rates were caused by nearly all tested compound groups, except the microorganisms and soap. Chrysopids were not harmed by acaricides, most pyrethroids, soap or microorganisms, but were affected by most of the insect growth regulators (IGRs) and most of the OPs. For predatory bugs, pyrethroids, carbamates, most OPs and few of the IGRs proved to be harmful. Fungicides and herbicides were relatively harmless for coccinellids, chrysopids and predatory bugs, but partly harmful to predatory mites.

- *Effects on beneficial parasitoids*: Synthetic pyrethroids and pyrethrin were very harmful to adults, regardless of the test species. In tests with the protected stages, several pyrethroids were only slightly harmful, but in combination with a persistence of more than one week this advantage was neutralized. OPs were very harmful to the unprotected stages and with few exceptions also to the protected life stages, and showed high persistence as residues. Carbamates were harmful in both types of laboratory tests, but some had a persistence shorter than three days. IGRs and most of the acaricides were harmless to both the susceptible and the protected developmental stage of the parasitoids. Plant extracts (except pyrethrin), soap and microorganisms were harmless. Fungicides belonging mainly to the group with a broad-spectrum and protective mode of action were harmful to adult parasitoids and revealed detrimental effects which persisted over one week. In tests with the protected life stage, however, all fungicides were considered harmless. Very few herbicides were harmful to adult wasps, but not for other developmental stages.
- *Entomopathogens*: Only a small number of carbamates out of the tested insecticides/acaricides affected entomopathogenic nematodes, while fungicides proved to be mainly harmless. Insecticides, acaricides and herbicides in most cases did not adversely influence the mycelial growth or the sporulation of the fungal species *V. lecanii, B. bassiana,* and *M. anisopliae* in laboratory tests or during infectivity tests in the greenhouse. Half of the fungicides examined in all types of tests affected at least one of the three test fungi, whereas one fourth of the fungicides were harmless for all of them. Effects could not generally be attributed to the mode of action of the fungicides. *Verticilium lecanii* was slightly more affected than *B. bassiana.*
- *Sublethal effects on natural enemies:* Besides direct toxicity caused by a number of "classical" insecticides, sublethal effects were also demonstrated in several investigations. Among sublethal effects of pesticide application on natural enemies which are reported in the literature are: development prolongation, reduced egg production or its total inhibition, decrease in prey consumption, changes in searching or foraging behaviour, alteration of pathogenicity in entomopathogens, and increased tendency to escape from treated surfaces. The importance of repellence of

pesticide compounds for beneilcials is difficult to classify. On the one hand, repellence may negatively influence natural enemies by expelling them from their host or prey which they need for further population development; on the other, beneficials can be protected from possibly hazardous contact with contaminated plant surfaces or prey/hosts. Both effects are undesirable, especially in greenhouses, where mass reared arthropods are intentionally introduced as biological control agents, and because the natural enemies would cease to be effective as control agents, particularly when untreated refuges are scarce. Insect growth regulators, like diflubenzuron, chlorfluazuron, fenoxycarb, flufenoxuron and teflubenzuron, which are incorrectly considered as harmless to many beneficials, in fact interfere with the viability of eggs, the moulting process, and the reproduction of several predators. The influence of different formulations of pesticides on their effects on natural enemies was shown for endosulfan, which as an emulsifiable concentrate (EC) formulation resulted in up to 17 per cent less mortality of *P. persimilis* than the wettable powder (WP) formulation in a residual laboratory test. For *E. formosa* the EC formulation of tebufenpyrad was more toxic than the WP formulation.

SOIL BIOTA AND NUTRIENT DYNAMICS UNDER ORGANIC SOIL MANAGEMENT

Soil biota is a functional component of soil systems, of the farm, and of the ecosystem. Pankhurst et al. clearly state the importance of soil biota: The soil biota is a highly diverse assemblage of organisms that carry out a wide range of processes that are important for the maintenance of soil fertility and soil quality. The development of sustainable farming systems will depend greatly on our ability to capture the benefits that may be derived from improved management of soil biota. This will only be achieved through an increased understanding of the soil biota, the functional processes it caries out and how soil and crop management practices affect its activity.

Organic farming systems emphasise living soil, and this includes the importance of soil organic matter, soil biota and food webs in the soil. Brought to general public attention by Rachel Carson's Silent Spring, our knowledge of the importance of food webs has grown rapidly, along with the understanding developed by ecologists such as Odum. The increasing interest in soil biology and ecology and soil quality means the relevance of food webs for agriculture and horticulture has been recognised.

One of the most profound aspects of the soil food web is the functioning of the rhizosphere. Rhizosphere soil generally has a higher number and broader diversity of soil micro-organisms, that is a more complex community structure, than the surrounding soil due to the interaction of plants with soil. Plant exudates and organic matter stimulate microbial activity and Ingham states that many plants can be recognised by their rhizosphere community structure. It appears that every plant species "co-creates" its own rhizosphere population or community structure that aids its nutrient uptake and can by various techniques protect it from potential pathogens. This means that a broad diversity of crops or plants has a more complex community structure of soil biota.

During the development of the soil system, the food web becomes increasingly complex. In the later stages of the soil and ecosystems development, the role of fungi increases. The same is true for biological activity. Young soils are low in biological activity (simple food web or chains). During the developmental stage biological activity increases (increased food web complexity), then activity stabilises (reaches a relatively steady state that fluctuates with daily and seasonal rhythms) during the climax or mature stage (complex but stable food webs).

As stated an increase in food-web complexity in an ecosystem goes together with:
A decrease in yield (or net community production);

1. (Semi) closed mineral cycles;
2. Increased turnover time and storage of essential elements;
3. Increased internal recycling;
4. Increased nutrient conservation;
5. Increasingly mutualistic behaviour;
6. Decreased entropy (or reduced losses).

The Primer recommends more research into community structures and food web complexity. Microbial biomass and microbial activity in the soil is a soil quality aspect and organic soil management has a positive influence on them. This might be due to an increase in food-web complexity. The hypothesis that the stability of soil ecosystems increases with increasing diversity (maturity) seems to be confirmed by Fleissbach et al. from FiBL. They suggest that the long-term effects of organic and conventional farming systems result in differences in the soil microbial community structure that lead to differences in the decomposition of organic matter. This results in higher or lower nutrient availability for the crops grown.

The introduction of micro-organisms to the farming system (often termed functional or effective, beneficial or plant growth promoting organisms) is being widely researched. In the USA and Europe compost and composted manures amongst other things are used to introduce and maintain soil organic matter and soil micro-organisms. In many Asian, African and Latin American countries effective micro-organisms are being trialled with promising results (*e.g.*,Sangakkara and Higa, 2000). Effective micro-organisms are cultured by fermenting rice-based organic material. Other "inoculums" like compost teas and fermented teas (herb, seaweed) can be used in a similar way. Biodynamic preparations are at times included in this list. Nutrient cycling and energy flows in terrestrial ecosystems are tied to the turnover of organic matter in soil. Although small in mass (we are still talking about tons per acre), the microbial biomass is amongst the most labile pools of organic matter and thus serves as an important and dynamic reservoir of plant nutrients. The succession of the soil microbial population during ecosystems development and soil development shows that a simple food web is vulnerable to external stress and largely based on a limited number of mainly bacterial species that build up a food source in the soil. Once the humus has reached a certain site specific level, the soil microbial population stabilises into a long-term energy-efficient, harmonious, (semi) closed system and supports a mature or climax ecosystem. The soil food web structure has changed from being relatively simple, vulnerable and bacterial based into a complex, stable community structure, rich in diversity and with an increased fungi population.

The first phase mainly represents growth and production, while the mature phase represents stability and quality. The mature system has an increased nutrient buffer in the form of higher (soil) organic matter and biomass content and a higher rate of internal recycling, meaning less nutrient losses from the system and more nutrients tied up in the biomass. With higher organic matter contents and complex soil food webs, soil borne pests and diseases are suppressed.

Soil biota is clearly important for (organic) soil management. It is a key aspect of plant nutrition, of pest and disease suppression, as well as of environmental degradation.

Organic farming is about finding the balance between the juvenile stage (high production, vulnerable to stress/perturbation/disturbance) and the mature stage (stability, quality, efficient). Mainstream farming emphasises the juvenile stage (high production, vulnerable to pests and diseases, more losses from the system). As a net result, the mature stage has a semi-closed nutrient cycling process, meaning lower losses from the system and increased nutrient sinks/pools. A clear example is the excessive nitrogen losses from the farming system.

The major water contaminant in North America and Europe is nitrate-N, which is also becoming an increasing challenge in New Zealand. The recognised principal sources of N-loss are the conversion of native to arable land use, fertilizers, animal manures and other nitrogen inputs. Soil management practices are known to influence N leaching. The major factor influencing N-losses are excessive rain or irrigation, mineralisation, immobilisation, nitrification, denitrification and humification. The net results of these processes generally result in less nitrogen leaching under organic management than under conventional management. Although not all research indicates a reduction in N leaching on organic farms, reduction of all losses from the system and thus reduction of N losses is one of the aims of organic farming systems. In 1993, an international workshop on nitrogen leaching in ecological agriculture identified key factors and recommended research work aimed at improving and designing organic farming systems. Environmental and soil remediation has become a major (knowledge) industry. The contribution of organic research centres to these challenges of reducing nutrient leaching and water contamination has been recognised. For instance, LBI has been developing a model for nitrogen dynamics for ecological arable farmers called NDICEA (Nitrogen Dynamics In Crop rotations in Ecological Agriculture).

Work on the model has increased knowledge of regional and seasonal mineralisation rates and net results of the competing processes of mineralisation, immobilisation and humification. NDICEA is becoming a promising instrument to assist in farm management and facilitation by researchers. More important, it has created insights into the relevant regional and seasonal soil processes. The model will be exposed to testing under various different (soil) conditions in further research programmes.

Models can be useful but have disadvantages too. Condron et al. reported on nutrient dynamics using budgetary models developed with mainstream farming research data. They expected that organic farming systems could support themselves with biological nitrogen, but expressed concern about continuing phosphorus availability. Nutrient budgets are a good linear approach but fail to explain what goes on inside the "Black Box", in this case the soil system, and the input-output approach does not explain the internal functioning of the system.

We know that the soil under organic soil management is generally different (see soil quality sub-section) from soil under conventional farming systems, which mean that the behaviour of the soil system under organic management is different. As a consequence, the current mainstream working models or budgets might not be appropriate tools for research on organic farming systems.

The results of this research support the hypothesis that the biodynamic system invokes higher efficiency of the soil microbial community with respect to substrate use for growth. In other words, they make better use of the soil's natural resources. The soil system ("Black Box") functioned more efficiently under organic farming conditions.

Waldon et al. also compared the performance of two very similar soil ecosystems under organic and conventional management. They concluded that the lower N and P contents of the organic site, considered with their healthy plants and high production, call for re-evaluation of soil test methods and standards for organic farming systems. The organic farming systems may use mineral nutrients in a more efficient manner and allow lower inputs.

This can partially be explained by the use of organic fertilizers as opposed to inorganic fertilizers. Moritsuka et al., (2001) studied the effects of organic and inorganic fertilizers on the nutrient dynamics in the rhizosphere, and concluded that the contribution of the net supply of N, P, and K through the replenishment from the solid phase was higher for the organic fertilizer treatment than for the inorganic fertilizer treatment. In short, inorganic fertilizers reduce the efficient use of the soil system's mineral resources.

In the light of these results, the concern of Condron et al. about continous phosphorus availability under organic soil management might not be warranted. Phosphorus availability is largely determined by mineralisation (by phosphatase enzymes) due to microbial and plant activities. Phosphorus mineralsation rates are correlated to nitrogen mineralisation rates, microbial activity and humus. It is decreased by high phosphate concentrations. Mineralisation rates are also higher in the rhizosphere.

Oberson et al. reported that the soil microbial biomass and the activity of the enzyme acid phosphatase were higher under the organic soil management plot in their long-term field trial. These results were attributed to both the higher quantity of organic carbon and organic phosphorus (organic matter) applied in these systems and also to the absence of or severe reduction in chemical plant protection. A further conclusion drawn was that phosphorus could not be the factor limiting crop yield under organic management. The ability of phosphorus to leave the soil solid phase was significantly higher under the biodynamic treatment than under all other treatments. This was explained by the higher calcium and organic matter contents in this system. Other important factors in increasing phosphorus uptake are the presence of mycorrhizal associations and earthworms, which have been extensively reviewed. The soil system with its components of soil organic matter, soil life community structure and nutrient dynamics plays an essential part in the global water, carbon, nitrogen, phosphorus and sulphur cycles. Soil organic matter and biomass are a major sink or pool for carbon, nitrogen, phosphorus and sulphur. The turnover, cycling rate and availability are under the influence of soil organisms. According to the Soil Biology Primer the food web serves the land manager in the following ways: fertilizer requirements may decline as a healthy food web efficiently stores and recycles nutrients;

- Nitrates do not leach into groundwater when soil organisms hold nitrogen in the rooting zone;
- Water quality is protected when organisms effectively degrade pollutants;
- More water soaks into the soil and can be used by crops as biological activity enhances soil structure;
- Pesticide use can be reduced as disease suppression improves with a healthy soil food web.

Landscape Quality and Organic Soil Management

Landscape quality, directly related to soil quality, is beginning to attract attention overseas but is underexplored in New Zealand. The degradation of the environment (soil, water, air), loss of nature (flora and fauna), and depletion of natural resources go hand in hand with the degradation of landscapes and loss of livelihood in rural communities. Agenda 21, the report of the United Nations Conference on Environment and Development has further stimulated national and regional authorities to take action, which has led to increased global interest in biodiversity and conservation and in sustainable and ecological development.

Promulgation of the Resource Management Act showed New Zealand as a front-runner in these developments. There have also been various fragmented efforts to enhance overall environmental quality in New Zealand. In Europe, concern about landscape quality led to the development of the EU Concerted Action group, which encouraged structured scientific research. The EU Concerted Action group developed a checklist for EU sustainable landscape management as a tool for policy making. Organic farming is increasingly recognised as being part of that development.

The basic philosophy of organic agriculture, allows the presumption that at least in potential, organic farming systems can contribute to the visible landscape quality in a positive way.

A database search or organic farming and landscape literature resulted mainly in the recovery of European oriented literature. The available literature on landscape and organic farming systems in New Zealand is limited. A well-designed and managed organic crop rotation has the potential to contribute towards improved sustainable development in the catchments area (or landscape).

Organic farming systems have characteristics such as increased biodiversity that enhance the landscape:

- 200 per cent to 2000 per cent more beetles and earth worms;
- 50-700 per cent more species of bees, butterflies, bumblebees and spiders;
- 130-700 per cent more herbs and non-crop species as compared to conventional farms (Austria, Germany, The Netherlands, and Switzerland); 30-800 per cent more birds reported in Germany, the Netherlands and the USA.

The increased biodiversity of organic farms in Europe includes species from the endangered species list. On these organic or ecological farms there can be an increase in diversity of land use types, including more use of woodlots, linear planting and crops compared with practices on traditional farms. Kuiper reported on the diversity of landscape elements, ecosystems and species.

These reports show that organic (including biodynamic) management tends to increase the diversity of fauna and flora considerably. As stated by van Mansfelt et al., (1998), increased biological diversity seems to go together with increased diversities of land use in general, and diversity of labour and spatial structures. This view is supported by agro-ecologists such as Altieri:

The reintroduction of a mosaic structure into agricultural landscape composed of woodlots, fencerows, hedgerows, wetlands, farmyards, etc, can lead to the creation of multiple habitats for reproduction, feeding, and sheltering for a number of beneficial species. This addition or enhancement of biodiversity can restore or improve community homeostasis.

Organic farming systems contribute well to landscape quality and are therefore supported by various governments in Europe. Stroeken et al. indicated that landscapes deliberately developed on farms (landscape production) can be studied. Regional identity (based on ecological and historical strengths of the region) became an overall goal, encouraged by the Dutch Government, for example, to stimulate local products and tourism.

Soil and landscape develop naturally together into unique diverse units that vary across continents and islands, providing global biodiversity and thus creating the global life support system. A change in flora and fauna communities on top of the soil goes together with a change of community structure in the soil, and will affect nutrient turnover rates (microbial biomass and activity). An increase in biodiversity generally accompanies improved stability in and on top of the soil. Organic farming systems contribute to this, although there is recognition in Europe of the challenges of increased diversity in the agro-ecosystem. Unless well designed, such diversity could lead to chaotic fragmentation, so research is concentrating on coherence within the farming system as a whole. The basic farming concept in biodynamic farming is referred to as managing the farm identity or farm individuality. The appropriateness of nature and landscape development to the organic farm identity has been described by a team from the Louis Bolk Institute:

For the concept of farm individuality to be used... in landscape planning, three issues need to be addressed:

1. A method to describe the farm individuality;
2. The people who live and work on the farm are part of the farm individuality, so they should participate in the (research) planning process; and
3. Landscape is perceived as a dynamic system and individuality is also a dynamic concept.

The LBI team presented a method designed for landscape-development planning at farm level, based on the concept of farm individuality and a Goethean-phenomenological approach. The method, which involves a participatory on-farm research methodology, can be characterized as a bottom-up rather than a top-down approach. It enables farmers to coooperate in landscape planning with all their ideas, feelings and future plans for their farm. The method is illustrated in work recently carried out on a Dutch organic farm, the 'Noorderhoeve'.

8

Guidelines of Pest Management

MONITORING SYSTEMS

Field monitoring is the key to an integrated pest management programme. When performed systematically on a quantitative basis, it provides the information needed to detect important changes in pest activity and disease incidence; evaluate pesticide performance; and establish seasonal records for evaluating and upgrading managerial decisions. When weather data and cultural practices that influence pest behaviour are considered, pest behaviour and populations can better be anticipated.

Monitoring systems may be limited to a single primary pest Such as the citrus rust mite, and to environmental conditions that influence rust mite behaviour or they may include all major pests and beneficials. Monitoring techniques vary depending on individual preferences and management intensity, but should provide a consistent, repeatable means to numerically measure pest activity. Follow monitoring procedures upon which the control action thresholds were designed.

Monitoring frequency depends on the pests' reproductive capacities and time needed under optimum conditions for populations to reach economic damage thresholds. Because of the reproductive potential of citrus rust mite, biweekly monitoring of susceptible fruit is required from petal fall until harvest or arrival of cool weather. Other pests generally require less frequent monitoring. Outside of the petal-fall to harvest interval, monthly monitoring is adequate for the citrus mite complex and other potential pest or disease situations.

Sampling Methods: Sample size, location and selection are important to any sample programme. Each citrus orchard is unique and requires individual monitoring; the first step in developing a monitoring system is to decide how many trees need to be examined and, within this parameter, how many fruit, leaf or soil samples need to be observed or collected for adequate representation and sampling reliability. Sampling may involve trees randomly selected for each monitoring period or permanently designated station trees. Trees should be dispersed sufficiently to achieve representative sampling of the entire orchard. This can be accomplished in most rectangular orchards by sampling in a diagonal direction. For square orchards, a Z-shaped pattern is appropriate.

The station tree concept involves selection of representative trees at the time of orchard establishment or at the beginning of each season. Selected trees are flagged, numbered and revisited each monitoring period throughout the season.

Random samples provide information on a larger proportion of trees within an orchard over time. Station trees more closely follow pest or disease dynamics on a few individual trees. Both systems have merit and, depending on the desired result and amount of time to be invested, a combination of both methods may prove most useful.

Systematic recording of data in the field is time consuming but necessary for clarity in review and interpretation of results. If field counts are considered impractical and some form of rating scale is employed (*e.g.*, trace, low, medium, high, very high), the assignment of numerical values, from 1-5 is acceptable. Organization of the data in summary form following field sampling allows for id review and assessment of individual orchard situations. Information gathered during the monitoring process aids in planning future pest management strategies.

CONTROL ACTION CRITERIA

Control action criteria can be defined as specific pest population levels or threshold levels that signal impending economic damage to the crop or trees. Ideally, each threshold value considers the pest's life cycle, its potential for rapid population increases under existing or anticipated environmental conditions, orchard history, the type of damage likely to occur in proportion to control costs, other stresses the trees are experiencing and other relevant factors.

Progress has been made in establishing threshold values for citrus rust mite, the major economic pest of Valley citrus, but standardized criteria for other common pests are lacking. Suggestions in this text for key pests are based on guidelines developed in other citrus production states and observational data recently accumulated by local IPM programmes. Action criteria in any pest management programme are flexible and should be altered to fit individual management conditions.

Monitoring Guidelines

These guidelines provide information on how to monitor specific pests and offer insight on what constitutes reason for control action. Pre-season monitoring suggestions are provided to encourage early season inventory of pest populations and disease potential in advance of post-bloom control effects. Early monitoring provides information helpful in selecting the most appropriate and cost-effective pesticides. This early information may be useful in making volume purchases of pesticide materials and for determining which orchards merit immediate attention.

Citrus Mite Complex

Field studies of rust mite behaviour have indicated that populations in excess of 50 mites per square centimetre (cm^2) on fruit for 2 weeks or more are needed for rind damage to become visible.

Also, rust mite populations on infested fruit may more than quadruple in a 2-week period under favourable conditions. These data support the need for a 2 week monitoring frequency for rust mite and are useful in establishing action thresholds.

As a general reference, a threshold value of 3 to 5 mites/cm^2, depending on optimal or suboptimal mite reproductive conditions, has been used successfully for several years in the local TAEX-IPM programme. Structured to allow the grower 2 weeks' lead time before significant damage Could occur, and provide a cushion for any sampling weakness, this threshold value applies to the following sampling regime:

- Sampling is from selected station trees, one tree per two acres.
- Six fruits sampled per station tree; two fruits on each of east, north and west quandrants, selected at random from partial shade areas.
- Mite counts taken from shaded side of fruit using a 14X hand lens, counting all live unites present in four separate lens fields per fruit.
- Each lens field is roughly 1/4 cm^2 in area, for a total of 1 cm^2 per fruit or 6 cm^2 sample per station tree.

To avoid rust mite management decisions being biased by two or three heavily infested fruits, consider the proportion of infested fruit. Also, consider mite distribution within trees or within orchards. Under this particular sampling regime, less than 20 per cent incidence of infested fruit or of station trees would ameliorate average rust mite density values, and an incidence of more than 20 per cent would substantiate or override them. The following criteria are utilized:

- Less than three mites/cm^2 and less than 20 per cent of sample fruit with mites-no action necessary.
- Three to five mites/cm^2 and 20 per cent of sample fruit with mites-borderline situation.
- More than five mites/cm^2 and more than 20 per cent of sample fruit with mites-control should be initiated.

A sampling technique involving different action criteria has been used successfully by growers for several years. It involves random sampling of a single fruit per tree on 50 randomly selected trees within an orchard. Rust mite counts are expressed in numbers per cm^2. Action criteria vary according to irrigation schedules and weather outlook, but generally action is indicated when average mite density exceeds 20 mites/cm^2 and/or 25 per cent of sample fruit are observed to be infested.

Factors to consider when making control decisions regarding rust mites include:

- Variety-given similar rust mite densities, oranges are downgraded less by mite damage than grapefruit.
- Lead time needed by producer to spray.
- Weather outlook and time of year-heaviest rainfall periods are May/June and August/September. Evaluate borderline mite densities during these periods in light of possible re-entry difficulties.
- Irrigation scheduling-orchards expected to need water within 10 to 14 days of an anticipated action threshold should be sprayed prior to irrigation.
- Miticide used in previous spray treatment-mite populations may rebound quickly following use of low residual materials.
- Orchard history-certain orchards, because of location, cultural practices, tree spacing or lack of uniformity in tree sizes and other reasons, consistently experience rapid re-infestation following control efforts.

Citrus red mites, first observed in 1980, are distributed throughout the Valley and have become a major citrus pest, particularly in the mid-Valley area between US 77 and US 281. Red mites feed on tender new leaves and populations increase with new growth flushes. Population levels decline during hot, dry July/August weather and resurge with the fall flush. Information about natural control agents in the Valley is not sufficient to determine when control action is most appropriate.

Control action guidelines for red mite on Texas citrus are incomplete pending further research on natural control agents and tree tolerance levels. Factors to consider in evaluating need for red mite control include:

- General tree vigore.
- Mite reproductive potential and environmental conditions.

Texas citrus mites do not require specific control action in most seasons if trees and fruit are protected from other mite and insect pests. Threshold levels vary according to tree vigour and should be determined by period during which heavy feeding is sustained rather than by specific mite numbers. Numbers exceeding 10 adult mites per leaf over a period of 30 days have been associated with winter defoliation of crown leaves.

Major considerations in Texas citrus mite management include:

- Tree vigour.
- Current and projected climatic conditions.
- Deleterious effect of fungicides applied for melanose or rust mite control on beneficial fungi.
- Effect of certain fall scalicide applications on over-wintering mite

populations (azinphos-methyl and methidathion have been implicated with increased mite populations).
- Time of year-predisposition of mite-damaged leaves to defoliation by strong, dry winds greatest during fall/winter months.

False spider mites' increase in the past was associated with the use of organophosphate pesticides to control other pests and situations involving use of zineb alone for control of rust mite. With the present trend to use specific miticides such as dicofol, chlorobenzilate and fenbutatin-oxide, primarily for rust mite control, recent false spider mite problems have been associated with orchards where aldicarb soil treatment replaced post-bloom and late spring sprays. In aldicarb-treated orchards where the need for rust mite control may not arise until late summer, start monitoring for false spider mite in July and continue through September.

Armored Scales: California red scale is a major pest of citrus in the Rio Grande Valley. It attacks all parts of the tree aboveground and is a common cause of limb dieback and tree death in new trees. Populations usually are kept in check by natural enemies, but tend to reach damaging levels during abnormally dry seasons on trees adjacent to dusty roads. A small parasitic wasp, Aphytis lingnanensis, provides effective control in most seasons if not restricted by dust or by application of broadspectrum insecticides. Begin monitoring for red scale in late spring to early summer to advantageously schedule a spray application during periods of high crawler activity or the early growth stages when scales are most susceptible to control. Orchards having a history of heavy red scale infestations the previous summer should be monitored early following warm winters. Monitoring always is required from July through October when colonization on fruit is easily seen and infestation levels are more easily determined.

To limit scale damage below tolerances set for federal grade standards, begin control action when 5 per cent of the fruit examined have 10 or more live scales. Examine at least 10 outside fruits on both east and west quandrants of sample trees to calculate infestation levels. A hand lens will help distinguish live scale insects from dead or parasitized scales, which appear partially loosened from the peel or exhibit holes cut by adult parasites as they exited the scale body. Estimates of scale parasites from leaf samples are less reliable than those based on fruit samples because of the difficulty in determining leaf age and when parasitism occurred.

Consider the following for monitoring and control of California red scale:
- Monitor new tree plantings frequently as they are subject to relatively rapid infestation.
- Inspect every young tree in the planting at least twice during summer and fall months.
- Natural control by scale parasites often is reduced during the heat of July and August.

- Increased incidence of spherical yellow crawlers encountered when searching fruit for rust mites signals impending infestations in bearing orchards.
- Oil sprays are least injurious of all scalicides to scale parasite populations.

Chaff scale is the most common widespread armored scale pest of Valley citrus. It attacks all aboveground parts of trees and can result in serious downgrading of fruit, but infestations rarely cause twig and limb dieback. Several species of parasites are known natural control agents.

Monitoring for chaff scale is suggested during pre-season and biweekly scouting for rust mite. Grade standards for fruit are important in deciding if summer or early fall scalicide application is warranted. Control action criteria are the same as described for California red scale.

Secondary Pests: Various minor citrus pests capable of causing economic damage under favourable circumstances require consideration during monitoring of major pests. Secondary pest eruptions usually result from loss of natural enemies through improper pesticide use, exposure to off-target pesticide drift or weather extremes. The two most common secondary pests are brown soft scale and citrus mealybug. Outbreaks of these pests occur most frequently and persistently near fields traditionally planted close to cotton and vegetable crops. Brown soft scale infestations generally develop during the summer and fall months. Early signs of infestation are easily overlooked during monitoring activities until blackening of leaves by sooty mold fungi becomes noticeable. Tree vigour is diminished by feeding activities and inhibition of photosynthesis by sooty mold growing on the leaves. When localized infestations are found, spot treatment of trees on an "as needed" basis may be appropriate to reduce costs and prevent possible deleterious effects on natural enemies.

Citrus mealybug is controlled most effectively with pesticides during early stages before protective wax secretions and sooty mold form. Monitoring activities must be concerned with identification and quantitative assessment of first and second stage larvae. Mature mealybugs are highly resistant to contact pesticides. Clusters of fruit-usually the higher quality, inner canopy fruit-are preferred by mealybugs. Heavy infestations interfere with fruit development and may cause fruit drop. Sooty mold is difficult to remove at the packing shed, often resulting in downgrading of fruit.

Mealybugs have several predatory and parasitic natural enemies. Severe outbreaks in orchards not normally affected by mealybugs usually are the result of pesticide-related reductions or other unfavourable circumstances affecting natural enemies.

Whitefly infestations (citrus whitefly, citrus blackfly, cloudy-winged whitefly or woolly whitefly) occurring in isolated situations also can be associated with disruption of natural control agents.

Texas leafcutting ants can cause serious damage to citrus, particularly In young plantings, so monitoring notes on location and severity of infestations will facilitate control efforts. The leafcutting ant poses a special problem in well-drained sandy loam soils because it may nest some distance from the orchards Repeated defoliation of trees reduces tree vigour, eventually killing young trees. Locate the nest by tracking at dawn or dusk or during periods of cloudy weather because aboveground activity ceases during direct sunlight.

The information given herein is for educational purposes only. Reference to commercial products or trade names is made with the understanding that no discrimination is intended and no endorsement by the Cooperative Extension Service is implied.

Texas Citrus Mite

Tetranychidae: Eutetranychus banksi (Banks)

Texas citrus mites (TCM) are 0.3 to 0.4 mm long and the eight-legged adults vary in colour from yellow to dark green, with characteristic dark internal food spots on each side down the back. Female mites are broadly oval and somewhat flattened. Males are more slender with longer legs and are triangular in shape. Larvae have six legs and are light yellow. Eggs are disc shaped and usually are laid near the midrib or branching veins. The life cycle can be very short under high temperatures, leading to very rapid population explosions. Unlike the citrus rust mite, the Texas citrus mite prefers low humidity conditions.

Most of the time, greater numbers are found on leaves on the south side of the tree, and in the top of the tree. They usually prefer to attack upper surfaces of leaves. Leaves damaged by TCM appear gray, show mesophyll collapse and may drop during periods of drying winds or when trees are stressed by lack of water or other factors.

Mites also appear on fruit but to a lesser degree than on leaves, and rarely cause noticeable feeding damage on fruit. Texas citrus mites usually are more prevalent in the May to July period, although they may increase rapidly at other times when conditions are favourable.

Insect and Mites Pests

Mites: Description and Biology

Mites represent the single most economically important group of Texas citrus pests. The citrus rust mite together with the spider mite complex, including citrus red mite and Texas citrus mite, are responsible for a major portion of pest losses and require the bulk of pesticides used on citrus in Texas. Mites feed on fruit and foliage by piercing individual plant cells and sucking out the contents, thus, destroying the cell. Their pest potential lies in their ability to develop rapidly and their tremendously great reproductive potential, allowing for enormous populations to develop in a short period of time.

Citrus Rust Mite

Eriophyidae: Phyllocoptruta oleivora (Ashmead)

The citrus rust mite (CRM) is considered the most important pest of Texas citrus. Rust mites are long, wedge-shaped and light yellow, measuring about 0.1 to 0.2 mm long, generally not visible to the unaided eye. The citrus rust mite feeds on the exposed surfaces of fruit, destroying the rind cells leading to the characteristic russeting of fruit.

This damage may result in reduced grade and size, thereby drastically reducing the marketability and value of the crop. Rust mites feed and develop year-round on citrus foliage and, when available, will develop on fruit. Rust mites may complete a generation in 7 to 10 days during the summer while development is drastically slowed or may halt during cool periods.

Early rust mite damage to fruit in late spring or early summer is evidenced by bronzing of the rind, a condition commonly called "buckskin" fruit. Severely affected fruits do not expand normally and frequently exhibit rind cracking. Damage occurring later in fruit development is evidenced by a darkening of the peel where heavy feeding took place, generally dark brown to black on green fruit and rust-brown on mature fruit. Similar symptoms of melanose disease on fruit, particularly grapefruit, may be confused with mite-induced russeting, but usually can be differentiated by noting peel texture and checking for presence of melanose pustules. Grapefruit generally are more susceptible than oranges to fruit damage by citrus rust mites.

High populations of CRM may reduce tree vigour. Affected leaves develop a dull green colour and show reduced turgidity. Rust mites pose a threat to the fruit crop from April to October, particularly during periods of warm, humid weather with frequent, light showers. If the latter conditions extend into early winter, fruit russeting may occur as late as December.

A reliable sign of impending rust mite damage to fruit is the dull, cloudy appearance of heavily infested fruit caused by thousands of mite skins cast off during the molting process. The distribution of citrus rust mites within trees varies during the year and is generally determined by environmental conditions. Rust mites tend to seek out high humidity areas away from direct sunlight but also avoid areas where dew forms. Rust mites overwinter on foliage and in bark crevasses. On foliage, mites are most likely to be found on undersides of dry, inner canopy leaves. Mites appear to prefer fruit on the west or northwest quadrant of trees during May to June when cloudy conditions prevail. As summer temperatures increase, mites move inward and toward the north quadrant, away from the direct sunlight. After harvest during winter months, highest mite populations have been observed on east or northeast quadrants.

PRINCIPLES OF PEST CONTROL

Pest control is the study of the development and spread of plant diseases and arthropod pests and the factors affecting these processes. The level of disease or pest infestation is the result of many interacting factors and this level determines the yield loss that the grower suffers from the pathogen or pest. In many respects, the methodology of research and the underlying concepts are very similar for bacterial and fungal plant pathogens, insects, mites, viruses and nematodes.

PEST TETRAHEDRON

The pest tetrahedron is used to envisage the interaction of diseases and pests with their environment. The tetrahedron consists of four components, which can all influence each other and together determine the level of the disease or pest. The four components are the plant pathogen/pest, the host plant, the environment, and human activity. Generally, it can be said that chemical control is only aimed at influencing the pest or pathogen directly, whereas integrated control may reduce the level of disease or pest by influencing any or several of the four components of the tetrahedron. A thorough knowledge on the influence of different factors on pests and diseases offers the basis for integrated control. The four components of the tetrahedron will be discussed below.

THE PATHOGEN/PEST

Infection Cycle of Plant Pathogens

The infection cycle of fungal plant pathogens consists of the following phases:

- Infection (germination of spores, penetration of plant tissue and colonization of plant tissue);
- Sporulation (production of spores and maturation of spores);
- Dissemination (spore liberation, spore dispersal and spore deposition).

Bacterial pathogens and viruses generally have a similar but simpler infection cycle. Some pathogens only complete one infection cycle per season, but most pathogens complete several or sometimes many cycles per season. The amount of disease that develops is the sum of the successes of the different phases in the cycle. All phases are subject to influences from the environment, the host plant and human activity.

Life Cycle of Pest Organisms

Although pest organisms belong to various taxonomic groups and consequently their life cycle varies accordingly, a general pattern for phytophagous arthropods may be described. Dispersal usually occurs at adult stage; adults look for an exploitable habitat and then a plant to feed and oviposit on. Once a plant is recognized as suitable for feeding and ovipositing, an adult female lays eggs (ovipary) or deposits nymphs or larvae (vivipary). Progeny commonly feed and develop on the plant where oviposition took place or on neighbouring host plants until reaching the mature stage. The adults may feed and oviposit on the same plant or move to younger and more suitable host plants. Many arthropods spend all their life on a plant but others, particularly among holometabolous insects, either have soil-inhabiting stages or the adults feed on a different host plant than the immatures did. Most greenhouse pests are multivoltine (several generations within a year), and univoltinism (one generation per year) is rather rare.

Research on disease epidemiology and pest population biology, can study the effect of different factors on the phases in the infection/pest cycle or can determine the effect on the resulting amount of disease or pest density directly. Studying the effect on each of the processes described – for instance germination and spore production in the infection cycle, or host plant selection and exploitation in a pest cycle – often yields a good understanding of the effect of a factor on a disease or amount of damage caused by a pest. This kind of research is best done under controlled conditions, which allows the fluctuation of only one or two factors. Research on epidemiology of a disease or pest population biology in a whole crop is usually based on monitoring many factors and analysing their respective effects through regression analysis. It is important that in such research all the relevant factors are monitored.

THE HOST PLANT

Traditionally, the host plant has been considered as a more important factor for the development of a disease than for determining the damage caused by an arthropod pest. This may be one of the reasons why plant resistance has been used more frequently to control diseases than to control pests. In fact, the tetrahedron scheme is common in plant pathology texts but rare in applied entomology books. However, this situation has changed recently as entomologists have intensified studies on insect-plant relationships. Different cultivars may vary in their susceptibility to certain diseases or pests, even if none of them is completely resistant.

This partial resistance may influence disease development by decreasing the number of successful infections by the pathogen, by increasing the latency period, by reducing the rate of lesion expansion or sporulation, or by any combination of these processes. The result will be a slower development of the epidemic. Similarly, different cultivars may influence the host plant selection by a pest and/or its oviposition, development and survival, and thus the rate of population increase. Many partially resistant cultivars express the same resistance during their entire life, but some resistance may depend on the physiological age of the plant.

Agerelated resistance can be either adult plant resistance or young plant resistance. On another scale, certain parts of the plant may be more or less susceptible to disease because of their age. The physiological status of the plant is affected by temperature, humidity and nutrition. Nutrient deficiencies or certain climatic conditions may predispose plants to the development of diseases and pests, but unbalanced fertilization may also increase this kind of problems. For example, excess of nitrogen amendments renders the plant more susceptible to *Botrytis cinerea* Pers.:Fr. and enhances the population increase of homopteran pests like aphids and whiteflies. A thorough knowledge of the factors affecting the incidence of diseases and pests through their host plant may help to prevent outbreaks by applying correct crop management practices. Because the host plant is such an important factor in disease epidemics and pest population dynamics, it is important to carefully monitor the host plant in any epidemiological research.

This means not only that cultivar and planting date should be recorded, but also plant spacing, nutrition, development stage of the plant and plant growth. Recording plant growth is also important for another reason: whether the disease is assessed as a percentage of total plant area – for example leaf area covered – or as a pest density – for example, the number of insects per leaf area – it is important to know the size of the host plant. Also, a count of the number of lesions for example may yield the same number for different crops, but if the number of leaves or other aspects of the size of the host plant are different, the impact of this number of lesions will be different and results difficult to compare between crops.

THE ENVIRONMENT

Many components of the environment can indirectly influence the severity of disease and pest injury through the host plant or by a direct effect on the pathogen and pest. Here, only the direct impact of the environment on the pathogen and pest population is dealt with. The main influences on pathogens stem from temperature, relative humidity (RH), radiation and wind. Pathogens are often affected by climatic conditions in most of the phases of the infection cycle. Germination of spores and superficial germ tube growth often show an optimum curve for temperature. For most pathogens, germination only occurs above a certain (high) level of RH or in the presence of wetness on the plant surface.

Lesion expansion is often influenced by air temperature, since this also affects the temperature of the plant tissue, but less by RH. Sporulation is affected by temperature and RH, whereas spore dispersal is often mostly influenced by air humidity and movement. It is important to realize that all these environmental factors will be different at different crop heights. It is therefore necessary to measure microclimatic conditions at a height where the pathogen is expected to attack the host plant. Manipulation of the environment by, for example, changing radiation with different covers, and/ or changing temperature, RH and ventilation by using different heating and/ or ventilation regimes, may influence diseases and pests. Concerning pests, their density is directly affected by many biotic and abiotic factors of the environment other than the host plant, such as competing herbivores, natural enemies and climatic factors. Among the latter, temperature, RH, light (quantity, quality and periodicity) and air movement are usually the most decisive in determining behaviour, phenology, survival, fecundity and dispersal of pests. All these abiotic factors may also affect natural enemy populations and, therefore, pests in an indirect, but important, way. Understanding all these complex interactions is crucial for detecting the most decisive factors that govern pest population dynamics and managing the environment for pest control accordingly.

HUMAN ACTIVITY

Humans can affect both the host plants, the greenhouse environment and the pathogen or pest organism by cultural practices and by application of chemical, biological and other types of control. Cultural practices may be aimed at rendering the plant less susceptible or acceptable to diseases and pests, or eliminating (or at least reducing) infection sources. Furthermore, any cultural practice changing plant growth, such as leaf area, will influence the microclimatic conditions for the pathogens and pest organisms in the crop. Greenhouse crops tend to be labour-intensive and this can lead to a more frequent spread of pathogens and pests over the crop by workers than in field crops.

Disease Epidemics and Pest Population Dynamics: Bases for Intervening in Agroecosystems to Reduce Losses

The amount of disease or the pest density resulting from the interactions between environment, plant and pathogen/pest, and the influence of humans on these factors, is the subject of plant disease epidemiology and pest population dynamics research. Plant disease epidemiologists and agricultural entomologists have approached these studies in seemingly different ways. However, given that both types of scientist deal with populations of living organisms that are subjected to common phenomena such as birth, death, development and migration, approaches of plant disease epidemiology and pest population dynamics are basically similar. Whereas the plant pathologist deals mainly with the effects of the harmful agent (the pathogen), that is the disease, the entomologist is more concerned with the agent itself (the pest). Consequently, plant pathologists usually measure the disease incidence or severity, whereas entomologists generally estimate the number of pest individuals.

DISEASE EPIDEMICS

The amount of disease changes over time. The curve of the amount of disease against time is called the disease progress curve (DPC). It is typically of a sigmoid shape, with a slow increase in the beginning, followed by a logistic increase and a levelling off at the maximum level of disease.

Vanderplank showed that DPCs can be described by logit:

$$(y_1) = \text{logit}(y_0) + r^* \, t$$

where logit(y) = ln [$y/(1-y)$], y_t = disease on time t, y_0 = initial disease on time 0, r = rate of increase, and t = time. The logit transformation will turn a sigmoid curve into a straight line, which enables an easier comparison of DPCs than the original sigmoid curves.

For some diseases, transformation by gompits is better than logits:

$$\text{gompit}(y) = -1n\ [-1n(y)]$$

Disease can be reduced by reducing initial inoculums (y_0) lowering the rate of increase of the disease (r) or limiting the duration of the epidemic (t) by delaying its start through preventive measures. Reduction of initial inoculum is the purpose of sanitation measures before planting the new crop, but inoculum can also be reduced during the epidemic by removing diseased plants or plant parts. Reduction of inoculum will delay the epidemic. It depends on the kind of pathogen whether this delay provides sufficient control. In the case of a pathogen whose spores are abundantly present in the environment, for example powdery mildew fungi, the reduction of initial inoculum will only give a small delay in disease progress. In the case of rare inocula, sanitation may provide almost complete control. Sanitation takes place both in the nurseries and in the greenhouses. Most manipulations of either the pathogen, the environment or host-plant susceptibility are aimed at a reduction of the rate of disease progress. This is achieved by slowing down any of the processes in the infection cycle.

PEST POPULATION DYNAMICS

Malthus' equation, initially developed to describe human population growth, was soon adopted by entomologists to study insect demography.

The equation predicts that a population will grow exponentially according to:

$$N_t = n_0 e^r$$

where N_t is the number of pest organisms at a specified time, N_0 is that number at an initial time (0), e is the base of Naperian logarithms, r is the rate of population increase, and t is the elapsed time. If r is assumed to be constant and independent of conditions that affect pest development, survival and reproduction, population growth is unlimited. This rate of increase r is also called intrinsic or maximal rate of increase and as such is referred to as r_m and depicts the rate at which the population would increase under permanently favourable conditions. In nature, however, favourable conditions are never indefinitely maintained and several – usually many – factors limit or retard population growth.

To reflect this, the so-called Verhulst' model predicts that populations will grow until reaching a maximum following a logistic or sigmoid curve that can be mathematically expressed by:

$$N_1 = K/\{1 + [(K-N_0)/N_0]e^{-r}\}$$

where *K*, called *carrying capacity* (the maximum population size that the environment's resources can sustain), is the asymptote of the sigmoid curve and the rest of elements as in the formula of exponential growth. The parameter *K* is a measure of the global effect of all environmental factors that limit the growth of a population, the so-called environmental resistance. The shape of population growth in Verhulst' model is represented in figure. Note that it can also represent the logistic model of Vanderplank and, although biologically unrealistic, it allows to show how pest control procedures may prevent pest populations to reach damaging densities. Like in disease control, pest population growth may be reduced by decreasing or delaying immigration of first pest individuals into crop plants (lowering N_0 or *t*), or by decreasing the rate of population increase via integrated enhancement of environmental resistance, for example by release of natural enemies (lowering *K*). In the chapter on plant resistance, the reader will find a discussion on how different effects on pest development and reproduction influences r_m.

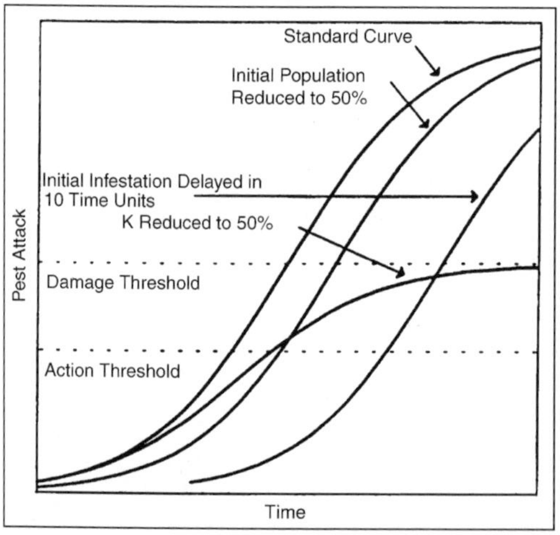

Fig. The Logistic Curve of Population Increase and the Different Procedures to Prevent Pest Population to Reach Damage Threshold. Initial Population can be Reduced (e.g. By Planting Uninfested seedlings) or K can be Lowered (e.g. Host-Plant Resistance or Biocontrol) or Initial Infestation Delayed (e.g. Sanitation).

Damage Relationships

Knowledge of the impact of diseases and pests on crop yield is needed for decisionmaking in disease/pest control. Decision-making is based on the damage relationship, that is, the level of yield loss associated with different amounts of disease or levels of pest attack. Yield loss can be either loss of quality or loss of quantity or both. Usually, damage relationships for a certain pathosystem or pest/plant interaction are expressed in quantity of yield, either in relative or absolute units. Most frequently, damage relationships are assessed empirically and analysed with regression analysis to estimate yield or yield loss from observations on the amount of infestation. The main drawback of this approach is that the resulting models are descriptive and do not take into account the physiological processes underlying the yield. Despite this, the descriptive models can be very useful in integrated control.

Following Campbell and Madden, five descriptive models to calculate the amount of disease and yield loss relationship may be distinguished:

- *The single-point models or critical-point models:* These models estimate yield loss by determining the amount of disease on one given moment, usually determined by the physiological status of the crop, for example the onset of flowering. Less frequently, these models have been developed with time variables, for example the number of disease-free days or the time until a certain level of disease is reached. Single-point models have been developed mainly for diseases in cereals. Their use is limited in crops in which yield accumulates over a longer period of time or harvesting takes place more than once, as for example in greenhouse vegetables.
- *Multiple-point models:* These models use several disease assessments to estimate yield loss. This type of model is most useful in situations where disease progress can be highly variable, depending on the host plant or the environment.
- *Integral models*: These models use the summed disease pressure over a specific period of crop growth which is relevant to yield. This is determined by calculating the area under the disease progress curve (AUDPC). These models can not distinguish between an early moderate epidemic and a more severe epidemic which starts later with the same AUDPC. This can be overcome by assigning weighting factors to the disease assessments made on different times or by incorporating another factor, for example the number of disease-free days, into the model.
- *Response surface models*: These models predict yield loss by using two different types of variables, for example disease severity and crop growth stage.
- *Synoptic models*: They are multivariate models that estimate yield loss by incorporating all independent variables in one equation.

Much of the conceptual framework to estimate the relationship between amount of disease and yield loss may be applied as well to damage relationship concerning arthropod pests. For decision-making purposes, a linear function of the yield response to insect infestation is generally assumed. In case the crop is able to compensate limited injury, there is a level of tolerance associated with low pest density. Crop tolerance to pest attack may be relatively high when pests injure the leaves of fruit vegetables like tomato, pepper, cucurbits or egg plants; often even 30–40 per cent of leaf injury does not result in yield reduction. Sigmoid yield responses to pest infestation are more difficult to fit, but logarithmic or power transformations may linearize the damage relationship. The consideration of more than one pest or disease and crop variables render complex polynomial relationships (the synoptic models mentioned above) which are difficult to interpret and to use in decision-making.

If a linear yield response may be assumed or derived, damage relationship can be "easily" found with field data as it has been mentioned above in single- and multiple-point models. When pests are multivoltine and their numbers are quite variable along the season, the use of insect*days instead of seasonal mean insect densities may be more meaningful as noted in the above mentioned integral models. Methods and techniques for this kind of studies may be found in Teng. A different approach to determining crop loss is the use of dynamic simulation models. Generally, a model for the development of the pest or disease is combined with a model for crop growth and production. This approach generates explanatory models, which are expected to have a greater predictive value than descriptive models. However, the development of simulation models requires a lot more basic information on the physiological processes and on the effect of environmental parameters on epidemics, and are therefore more difficult to develop than models based on regression analysis.

Damage and Action Thresholds

In combination with the damage relationship for a given pathosystem or pest/crop interaction, it is important to assess the damage and action thresholds. The damage threshold is the maximum level of disease or pest attack below which yield losses do not occur. The action threshold is the level of disease or pest attack at which control action should be taken to prevent the epidemic or pest to reach the damage threshold. Because there are often no fully effective techniques to control a disease or a pest immediately, the action threshold is lower than the damage threshold. Generally speaking, action thresholds for diseases lie below the logistic increase in the DPC. Damage and action thresholds are an important tool in integrated control when several control alternatives are available.

Whereas the damage threshold essentially depends on the disease/pest level and yield loss relationship, the action threshold may greatly vary according to the efficacy of each of the control alternatives and how long they take to be effective. Action threshold to control a disease – or a pest – will be probably higher if we choose a quick acting pesticide than with biological control, where natural enemies need time to react. For example, the action threshold for greenhouse whitefly control may be up several adults per leaf if we rely on insecticides, whereas it is around one adult per leaf when the parasitoid *Encarsia formosa* Gahan must be used in seasonal inoculative releases for the biological control of the whitefly. The knowledge and application of action thresholds generally reduce the amount of control inputs compared to general practice.

Determination of the Thresholds is Not Always Easy

Yield loss can be defined as loss in weight of the harvested product or the loss in economic value. This potential economic loss can be compared to the cost of a control measure. The translation of yield loss in weight to economic yield loss depends on the expected price of the harvested product and is therefore difficult to perform. In greenhouse crops that are harvested continuously, this is further complicated by the fluctuating prices within one growing season. For example, one kilo yield loss in cucumber or tomato in The Netherlands will be much more costly for the grower at the beginning (early spring) and end (late fall) of the season when prices are higher than in summer.

Further complexity in the determination of the action thresholds leads to some other considerations:

- Long term consequences of the current decisions for the disease or pest levels (instead of considering just one generation or infection cycle);
- Influence of control actions on crop revenue (it may be different if the decision is made at farmer or regional level);
- The risk attitude of the grower. Regarding the latter, stochastic models (in which an occurrence probability is associated to each decision) are more useful thresholds for quantifying risks than the deterministic damage and action model.

Damage Relationships and Thresholds in Greenhouse Crops

Despite the importance of knowing damage relationships and damage and action thresholds for integrated control, very limited specific information is available for most greenhouse diseases and pests. This can be explained in flower and pot plant crops for the extremely low tolerance of their aesthetic value to the most common diseases and pests. When known, damage thresholds in ornamental crops are near zero, as in the case of powdery mildew in roses, where the damage threshold is only 5 pustules/m^2. The same pest may have quite different damage thresholds if vegetable or ornamental crops are considered.

For example, tomato may tolerate relatively high leafminer infestation (e.g. several dozens of mines per plant) with no yield loss, whereas 1–2 mines on chrysanthemum leads to cosmetic damage. Some thresholds are available for greenhouse vegetable diseases and pests. Currently, the damage relationship for powdery mildew fungi in greenhouse vegetables is determined in The Netherlands. For cucumber, the best fitting model to describe the damage relationship for powdery mildew caused by *Sphaerotheca fusca* (Fr.) Blumer was an integral model using AUDPC. The slope of the regression line between yield and AUDPC was similar in several seasons and for different cultivars.

In this case, early disease was also compared to late, more severe disease. Similar AUDPC values and similar yield losses were found for these situations, so the inability of integral models to distinguish between early and late epidemics does not seem to be very important here. An additional problem for the determination of damage relationships is the fact that some pests can inflict more than one type of damage concurrently. This is the case of greenhouse whitefly, that feeds on phloem sap, with the consequent debilitating effect on the plant, but it also damages leaves and fruits by producing honeydew on which sooty mould develops, hampering photosynthesis and respiration and rendering fruits unmarketable. Damage thresholds may be quite different depending on which type of damage is considered to first occur as whitefly populations increase and this is decisively influenced by RH.

In very humid greenhouse environments – particularly common in northern Europe – damage by sooty mould development occur at whitefly densities lower than those needed for damage derived from whitefly feeding activity. A density of 2500 greenhouse whitefly nymphs per leaf has been reported to cause yield reduction on tomatoes as direct consequence of phloem extraction, whereas a much lower density of 60 nymphs/leaf has been observed to produce sufficient honeydew to induce sooty mould development on tomatoes if RH reaches at least 80 per cent for eight hours. For *Bemisia* species, a third type of damage is known which is that related to their ability to transmit tomato yellow leaf curl virus (TYLCV). This adds even more complexity to the pest density and yield loss relationship.

The damage threshold is highly dependent on the amount of virus inoculum present in or near the greenhouse. Examples of damage thresholds for thrips pests in vegetable crops may illustrate the variability of values that can be found in the literature. For *Thrips palmi* Karny on aubergine the damage threshold is 0.08 individuals/leaf, whereas the same author gives a value of 4.4 thrips/leaf for cucumber. For sweet pepper, Kawai gives, for the same pest, a threshold of 0.11 thrips/flower. There is unlikely to be a single damage threshold for a given pest on a given crop, but many, depending on the market and climatic conditions.

Even more variable are the action thresholds of greenhouse pests in various IPM programmes implemented in the world. If natural control must be considered in biological control of greenhouse pests, abundance of naturally occurring predators and parasitoids is an additional element to monitor and consider in decision-making. This is the case of action thresholds to release

Diglyphus isaea (Walker) in Mediterranean greenhouses for the control of leafminers; the standard action threshold can be lowered if the parasitoid, that occurs naturally in the area, comes into greenhouses and establishes early in the season.

NATURAL PEST MANAGEMENT

Whether conventional or organic, all farmers are concerned with pests. They spend a lot of time and resources controlling them. However, in the organic "world view," pests — whether weeds, insects or diseases — are not simply scourges. They are indicators of how far a production system has strayed from the natural ecosystems it should imitate. Certain weeds, for example, tend to predominate when soils are too acidic or too basic; some become a problem when soil structure is poor and conditions become anaerobic; others may be stimulated by excessive fertilizer or manure salts.

Organic proponents also believe that insect pests are attracted to inferior or weak plants — the result of poor crop nutrition. Their logic continues by asserting that pests are naturally repelled by vigorous, well-nourished plants. This belief is often challenged, and significant research remains to be done. As scientific understanding has grown, insect pest outbreaks are also being understood as imbalances in the whole agro-ecosystem and how it is managed. In nature, massive pest outbreaks are relatively rare and short-lived, due to the presence of natural predators, parasites, and disease agents that quickly knock the pest numbers back down to a moderate level. In farming systems that inadvertently destroy or otherwise fail to support the natural control complex, pest problems are routine and, typically, worsen with time. The farmer becomes increasingly addicted to costly and extreme control methods to produce a crop. Most organic growers consider pesticides to be a cause of agro-ecosystem imbalances and employ allowed natural pesticides as little as possible.

Integrity refers to the systems in place and actions undertaken to assure that consumers of organic products get what they pay for. Consumers have a right to expect that the organic food they buy not only be raised by organic methods but be protected from contamination and from commingling with non-organic products.

While the responsibility for much of this rests with others in the organic marketing chain, many certified organic growers need to incorporate additional practices that work to assure the integrity of their products. Proper record keeping is very important in this regard, though growers are often reluctant to spend much time on it. Among the more important production practices in the field are buffer strips, which reduce chemical drift from neighbouring fields and roadsides, while also serving water and soil conservation objectives. The tools and practices of organic agriculture include traditional alternatives — crop rotation, manuring, liming, etc.—long recognized as important to a sound production system. They also include more contemporary practices and materials that research and keen observation over time have contributed. The following list of tools and practices is not intended to be comprehensive, though the primary options are addressed. Note, too, that each farm operation will employ its own combination of tools and practices to build a working organic system. There is no simple cookbook formula for combining them in ideal proportions.

CROPPING SYSTEMS OF CROP ROTATION

Essentially a tool for annual cropping systems, crop rotation refers to the sequence of crops and cover crops grown on a specific field. Particular sequences confer particular benefits to long and short-term soil fertility, and to pest management.

Agronomic operations are especially dependent on crop rotations that include forage legumes. These provide the vast majority of the nitrogen required by subsequent crops like corn, which is a heavy consumer of that nutrient. Even when livestock are present to generate manure, the animals are largely recycling the nitrogen originally fixed by legumes in the system. An example of a basic agronomic rotation, typical of that found on midwestern organic farms.

The basic Midwestern rotation demonstrates the elegant way in which a whole farm system can function:

- Legumes fix nitrogen in the soil, providing for subsequent non-legumes in the rotation.
- Several insect pest cycles are interrupted, especially those of the northern and western root-worm species that can be devastating to corn.
- Several plant diseases are suppressed, in cluding soybean cyst nematode.
- Weed control is enhanced as perennial weeds are destroyed through cultivation of annual grains; most annual weeds are smothered or eliminated by mowing when alfalfa is in production.
- Livestock manures (if available) are applied just in advance of corn, a heavy nitrogen consumer.
- All crops can be marketed either as is or fed to livestock to be converted into value-added milk, meat or other livestock products.

Ralph and Rita Engelken, widely respected organic pioneers in the 1970s and 1980s, used a similar rotation that suited their hilly northeast Iowa farm and supported their main livestock enterprise, *backgrounding* beef cattle. (Backgrounding is confined or semi-confined feeding of young range stock to increase their size before final finishing in a feedlot.) The feed ration the Engelkens relied on consisted mostly of haylage, corn silage, and ground ear corn.

MANURING AND COMPOSTING

Livestock manures are the most traditional and widely recognized organic fertilizers. Under ideal circumstances, livestock enterprises are integrated into the whole farm operation, and manuring becomes part of a closed system of nutrient recycling. This is still strongly encouraged in organic operations. In reality, however, crops and livestock production are often divorced from each other, and manures must be imported.

This has created some concerns in the organic community, since much manure is now generated by large, industrial agriculture feeding operations called CAFOs (Confined Animal Feeding Operations). Not only are there concerns about contaminants (heavy metals, antibiotics, pesticides, hormones) but many in the organic community also object to any "partnering" with this segment of conventional agriculture, which is considered at odds with the environmental and social values represented by organic farming.

Nonetheless, the National Organic Programme does not differentiate between CAFO and other livestock manure sources. However, the NOP regulations do require that livestock manure not contain any synthetic substances not included on the National List of synthetic substances allowed for use in organic crop production. Another issue that has grown up around manure use in organic farming relates to food safety.

At a time when concerns about microbial contamination are high, there are questions about the risks associated with manure use on food crops. A focus piece on the February 2000 television news programme 20/20 was especially controversial. The segment suggested that organic foods were more dangerous than other food products in the marketplace due to manure fertilization. The reporter ignored the fact that conventional farms also use manures. Were all the manure generated annually in the U.S. (about 1.4 billion tons) applied only to organic farm acreage (estimated at roughly 1.5 million acres in 1997), each acre would receive about 933 tons.

Furthermore, certified organic producers have strict guidelines to follow in handling and applying manures. The National Organic Programme regulations require raw animal manure be incorporated into the soil not less than 120 days prior to the harvest of a product whose edible portion has direct contact with the soil surface or soil particles, and be incorporated into the soil not less than 90 days prior to harvest of a product whose edible portion does not have direct contact with the soil surface or soil particles.

One of the best means of handling manures is composting. Composting stabilizes the nutrients in manure, builds populations of beneficial organisms, and has a highly beneficial effect on soils and crops. Compost can be produced on-farm by a number of means. Additional products from composts, such as compost teas, have special applications in organic agriculture.

Human manures are expressly forbidden in certified organic production. This includes composted sewage sludge (also called "biosolids"). The organic community made its opinion on this quite clear when the USDA's first draft of the national rule proposed allowing the use of sludge in certified production. It was counted as one of the "big three" targets of protest, along with food irradiation and genetic engineering. The prohibition of biosolids would have been disconcerting to Albert Howard, who decried the failure of cities to return their organic wastes to the countryside. Such recycling was, in his mind, a key aspect of sustainability.

GREEN MANURES AND COVER CROPS

Green manuring consists of incorporating into the soil a crop grown for the purposes of soil improvement. It is a practice with a long history. Green manuring has been ignored in recent years as a serious option for soil improvement because the traditional practice entailed planting a full-season cover crop. This removed the field from commercial production for a whole season. Interest has returned, however, since green manuring strategies have been combined with cover cropping schemes.

Cover cropping is growing a crop for the purpose of soil and nutrient conservation. It is a more contemporary concept than green manuring, in crop agriculture. The two concepts — cover cropping and green manuring — go well together, as most cover crops are easily used as green manures prior to the planting of a commercial crop. The combined benefits become economically feasible when the cover is grown during the off-season or inter-seeded with the main crop. It is made even more desirable when the cover crop includes nitrogen-fixing legumes.

LIVESTOCK ON ORGANIC FARMS

Among the thorniest of issues swirling around the edges of organic agriculture is the role of livestock. The disagreements arise because of the diversity of people and philosophies in the organic community. Organic agriculture can usually count vegetarians and animal welfare proponents among its more vocal supporters. Many of these people feel strongly that animals should not be exploited. Their rationale often goes beyond emotional and religious beliefs; convincing human health concerns, social issues, and environmental reasons are commonly cited. On the other side of this argument are those who feel that an organic farm cannot achieve its full potential or ecological balance without livestock manure; that it is essential to nutrient cycling and to the finer aspects of soil building.

Excellent soil fertility can be built in the absence of farm livestock and livestock manures by using vegetation-based composts and by harnessing the livestock in the soil— earthworms and other soil organisms. However, it is clearly easier to design a contemporary, low-input organic farm when traditional livestock are integrated. The biological and enterprise diversity that livestock can bring contributes enormously to stability and sustainability.

A good example is provided by Rivendell Gardens in Arkansas, which began integrating livestock enterprises after several years as a solely horticultural operation. The owners of Rivendell, Gordon and Susan Watkins, now rotate their strawberry and vegetable crops with grass-fed, direct-marketed beef and postured poultry. Ideally, poultry follows beef on pasture to reduce cattle parasites. The seasons in mixed legume/grass pasture leave the soil quite mellow and well-mannered for subsequent high-dollar horticultural crops.

The Rivendell operation demonstrates the sort of organic management where a large number of organic farmers and many animal welfare proponents find common ground. The Watkins' animals are all raised with minimal confinement and generous access to sunshine, fresh air, and free-choice foodstuffs. While domesticated and destined for slaughter, they lead low-stress lives in conditions much closer to natural than the conventional alternatives. This is the antithesis of industrialized factory farming systems, which are increasingly becoming the norm in livestock production.

Many in both the organic and animal welfare communities are working to prohibit factory farming of livestock in organic systems. Many of the difficulties revolve around the fine interpretations of language in various organic standards. Wording such as "access to fresh air and sunlight," for example, can be construed to mean nothing more than opening the door on one end of a large confinement poultry house for a couple hours a day.

What Howard had not taken into account is the almost universal contamination of urban wastes with heavy metals and chemicals that are not eliminated by composting and may even be concentrated. Perhaps this was not yet a serious problem in his time; it is, however, in ours. Organic farmers and consumers concerned about contamination of soil and crops with agricultural pesticides and synthetic fertilizers would be remiss to ignore the contamination hazards of even well-composted sewage.

Fred Kirschenmann, a farmer and former NOSB member, has written eloquently about the progress of the National Organic Programme. In a critique of the March 2000 draft of the proposed rule, he pointed to another reason why the use of biosolids ought to be prohibited in organic production. Because of the manner in which biosolids are generated, they are easily hauled and land-applied on an industrial-scale to industrial scale organic farms. Furthermore, since biosolids can essentially supplant animal manures as a source of organic matter and nutrients, their use would allow some very large farms to circumvent the traditional practices that promote Biodiversity and enterprise diversity and integration. What Kirschenmann fears from biosolid use is technology that would nudge organic agriculture down the same road of industrialization taken by conventional agriculture.

INTER-CROPPING AND COMPANION PLANTING

Interplanting two or more mutually beneficial crops in close proximity is one strategy for increasing biodiversity. In large-scale mechanized crop culture, this is called *Inter-cropping*. It typically involves alternating rows or a number of rows of compatible field crops, like soybeans and corn. It also applies to sowing multiple forage crops, like alfalfa, bromegrass, and timothy, when these are grown together.

When interplanting is done on a smaller scale, it is often called *companion planting*. A classic example of companion planting is the inter-planting of corn with pole beans and vining squash or pumpkins. In this system, the beans provide nitrogen; the corn provides support for the beans and a "screen" against squash vine borer; the vining squash provides a weed suppressive canopy and discourages corn-hungry raccoons.

Index

A

B

C

D

E

F

G

H

I

L

M

N

O

P

Q

R

S

T

V